SpringerBriefs in Computer Science

SpringerBriefs present concise summaries of cutting-edge research and practical applications across a wide spectrum of fields. Featuring compact volumes of 50 to 125 pages, the series covers a range of content from professional to academic.

Typical topics might include:

- A timely report of state-of-the art analytical techniques
- A bridge between new research results, as published in journal articles, and a contextual literature review
- A snapshot of a hot or emerging topic
- An in-depth case study or clinical example
- A presentation of core concepts that students must understand in order to make independent contributions

Briefs allow authors to present their ideas and readers to absorb them with minimal time investment. Briefs will be published as part of Springer's eBook collection, with millions of users worldwide. In addition, Briefs will be available for individual print and electronic purchase. Briefs are characterized by fast, global electronic dissemination, standard publishing contracts, easy-to-use manuscript preparation and formatting guidelines, and expedited production schedules. We aim for publication 8–12 weeks after acceptance. Both solicited and unsolicited manuscripts are considered for publication in this series.

More information about this series at http://www.springer.com/series/10028

Dániel Leitold · Ágnes Vathy-Fogarassy ·
János Abonyi

Network-Based Analysis
of Dynamical Systems

Methods for Controllability and Observability
Analysis, and Optimal Sensor Placement

Dániel Leitold 🅳
Department of Computer Science
and Systems Technology
University of Pannonia
Veszprém, Hungary

Ágnes Vathy-Fogarassy 🅳
Department of Computer Science
and Systems Technology
University of Pannonia
Veszprém, Hungary

János Abonyi 🅳
MTA-PE Lendület Complex Systems
Monitoring Research Group
University of Pannonia
Veszprém, Hungary

ISSN 2191-5768   ISSN 2191-5776 (electronic)
SpringerBriefs in Computer Science
ISBN 978-3-030-36471-7   ISBN 978-3-030-36472-4 (eBook)
https://doi.org/10.1007/978-3-030-36472-4

This Springer imprint is published by the registered company Springer Nature Switzerland AG
The registered company address is: Gewerbestrasse 11, 6330 Cham, Switzerland

Preface

Network science for the analysis of dynamical systems has become a widely applied methodology over the past decade. Representing the system as a mathematical graph enables several network-based methods to be applied and centrality and clustering measures calculated to characterise and describe the behaviour of dynamical systems. Through the application of network-related algorithms in complex interconnected systems, the methodology created a strong connection between computer science, control theory and other fields where the approach is applicable, e.g. in biology.

The key idea of this book is that the dynamical properties of complex systems can be determined by the effective calculation of specific structural features by network science-based analysis. Furthermore, certain dynamical behaviours can originate from the existence of specific motifs in the network representation or can be determined by network segmentation. Although the applicability of the methods and results was highlighted in the operability analysis of dynamical systems, the proposed methods can be utilised in various fields that will be mentioned at the end of each chapter.

Chapter 1 of this book provides a brief introduction to the methodology that utilises the network science-based representation of dynamical systems to determine the location of the inputs and outputs such that their cardinality is also minimised. Then the scientific impact and main applications, e.g. in biology, sociology or physics, of the methodology are expounded on.

Chapter 2 deals with the structural analysis of the network-based representation of dynamical systems and defines those motifs that are specified by the dynamics of networks, e.g. integrating behaviour of a state variable, the diffusion of communication or changes in a temporal network. It was pointed out that these motifs are usually missing from benchmark networks. The significant reduction in the number of necessary inputs and outputs according to the presence of these motifs is also presented.

Chapter 3 deals with the improvement of the input and output configurations of dynamical systems, and presents five methods that can be used to decrease their relative degree according to the revealed motifs of the networks.

Chapter 4 deals with the analysis of the correlation between the structural properties of network-based representations and dynamical behaviours of dynamical systems. A general workflow is proposed, which was presented through a case study, that involved more than 600 heat exchanger networks.

Finally, an Octave- and MATLAB-compatible toolbox is introduced which is capable of conducting the analysis introduced in the previous chapters. The implemented functions are presented in the form of an artificial example that clearly shows the connection between the structural as well as dynamical features and highlights the differences between the centrality measures.

The applicability of the developed algorithms is presented in the form of well-known benchmark examples. An in-depth knowledge of the benchmark examples is not required from the reader, however, this may be beneficial to understand the proposed approaches. The reader is encouraged to unleash his/her imagination in terms of the applicability of the methods in other fields since the problems and results can be transferred to other networks and system classes. The introduction of the proposed algorithms is supported by in excess of 50 figures and more than 170 references that provide a good overview of the current state of the network science-based analysis of dynamical systems and suggest further reading material for researchers and students. The files of the proposed toolbox are downloadable from the website of the authors (www.abonyilab.com).

We hope, with the help of this book, that computer scientists can take on new challenges as further algorithms are necessary to reveal all the information that is hidden behind the structure of the networks. What is more, control engineers will acquaint themselves with a new tool applicable to the analysis of interconnected and structurally complex systems. Additionally, other experts, e.g. biologists, physicists, pharmacists, doctors and logisticians, will receive an overview of how the analysis of the dynamic-related properties of networks can support their analysis.

This research was supported by the National Research, Development and Innovation Office NKFIH, through the project OTKA-116674 (Process mining and deep learning in the natural sciences and process development) and the EFOP-3.6.1-16-2016-00015 Smart Specialization Strategy (S3) Comprehensive Institutional Development Program.

Veszprém, Hungary   Dániel Leitold
September 2019   Ágnes Vathy-Fogarassy
                                                                    János Abonyi

Contents

Symbols and Acronyms

Symbols

State-Space Representation

x	Vector of state variables
u	Vector of inputs
d	Vector of disturbances
y	Vector of outputs
$\mathbf{A}$	State-transition matrix
$\mathbf{B}$	Input matrix
$\mathbf{E}$	Disturbance matrix (linear)
$\mathbf{C}$	Output matrix
$\mathbf{D}$	Feedthrough (or feedforward) matrix
b^j	jth column of matrix $\mathbf{B}$
c_j	jth row of matrix $\mathbf{C}$
$\mathcal{C}$	Controllability matrix
$\mathcal{O}$	Observability matrix
$\mathbf{I}_N$	Identity matrix of size $N \times N$
N	Number of state variables
M	Number of inputs
K	Number of outputs
f	State-transition function
t	Time

Network Representations

$G(V, E)$	Network representation of the system
G^u	Undirected representation of G

V	Set of vertices of the network representation of the system
E	Connections between the state variables
V_i	Node i from the node set V
E_i	Edge i from the edge set E
G_{SS}	SS-based graph
V_{SS}	Set of nodes of SS-based representation
E_{SS}	Set of edges of SS-based representation
G_{SM}	SM-based graph
V_{SM}	Set of nodes of SM-based representation
E_{SM}	Set of edges of SM-based representation
G_{DAE}	DAE-based graph
V_{DAE}	Set of nodes of DAE-based representation
E_{DAE}	Set of edges of DAE-based representation
V_{hs}	Set of nodes of hot streams in SM-based representation
V_{cs}	Set of nodes of cold streams in SM-based representation
V_{hp}	Node of high pressure steam in SM-based representation
V_{cw}	Node of cold water in SM-based representation
E_{he}	Set of edges of heat exchangers in SM-based representation
E_{uh}	Set of edges of utility heaters in SM-based representation
E_{uc}	Set of edges of utility coolers in SM-based representation

Configurations Extension

$N_R(i)$	Set of reachable nodes from node i
Z_i	Set of nodes to which the nearest sensor node is node i
W_i	Set of reachable nodes from node i in r_{max} steps
$\mathcal{U}$	Set of all state variables
C	Set of necessary driver nodes for structural controllability
O	Set of necessary sensor nodes for structural observability
J	Set of driver/sensor nodes of the input/output configuration
P	Set of state variables covered by set J
$Cc(i)$	Closeness centrality of node i
$Bc(i)$	Betweenness centrality of node i
σ_{st}	Number of shortest paths from node s to t
$\sigma_{st}(i)$	Number of shortest paths from node s to t that intercept node i
K^+	Number of additional sensor nodes
$\mathcal{L}$	Lie derivative
$r_{i,j}$	Relative degree of y_i and u_j
r_i	Relative degree of y_i
r	Relative order of the system
r_{max}	Upper bound for r
S	Set of sensor nodes

S_f	Set of fixed sensor nodes
S_c	Set of candidate sensor nodes
β	Weighting parameter of cost function
$M*$	Set of matched nodes
Δ	Difference between cost functions in SA
T_{max}, T_{min}	Maximum and minimum temperature in SA
m_{max}, m_{min}	Maximum and minimum fuzzy exponent in SA
T^i	Temperature in iteration i
m^i	Fuzzy exponent in iteration i
α	Reduction rate of temperature
α_m	Reduction rate of the fuzzy exponent
$maxiter$	Number of iterations of SA
$R(s_j)$	Set of the state variables of the cluster of sensor node s_j
$\mu_{i,j}$	Fuzzy membership function of x_j to y_i

Network Measures

n_i	Weight of node i
w_i	Weight of edge i
k_i^{out}, k_i^{in}, k_i	Out-, in-, and simple degree of node i
$\langle k_i \rangle$	Average node degree
ℓ_{ij}	Length of shortest path between nodes i and j
$\mathcal{D}$	Distance matrix
$TWC(G)$	Total walk count measure of graph G
$CNC(G)$	Coefficient of network complexity of graph G
$Ecc(i), Ecc(G)$	Eccentricity of node i, or graph G
$W(G)$	Wiener index of graph G
$AD(G)$	A/D index of graph G
$BJ(G)$	Balaban-J index of graph G
$CI(G)$	Connectivity index of graph G
d_{max}	Diameter of the network

Heat Exchanger Networks

T_{hi}	Hot input temperature of a heat exchanger
T_{ci}	Cold input temperature of a heat exchanger
T_{ho}	Hot output temperature of a heat exchanger
T_{co}	Cold output temperature of a heat exchanger
v_h	Flow rate of hot stream
v_c	Flow rate of cold stream

V_h Volume of hot side of heat exchanger
V_c Volume of cold side of heat exchanger
U Heat transfer coefficient
A Heat transfer area
c_{ph} Heat capacity of hot stream
c_{pc} Heat capacity of cold stream
ρ_h Density of hot stream
ρ_c Density of cold stream
U_{un} Number of units
N_s Number of streams
L Number of independent loops
S Number of separate components
H Number of heat exchangers

Water Tanks

A_i Area of water tank i
K_v State of valve i (open/close)
F_i Flow rate of valve i
x_i Water level in tank i

Acronyms

ARR   Analytical redundancy relation
CFD   Computational fluid dynamics
CLASA   Clustering large applications based on simulated annealing
CNC   Coefficient of network complexity
CO   Combinatorial optimisation
DAE   Differential-algebraic system of equations
DN   Distribution network
DRSA   Dual representation simulated annealing
FDI   Fault detection and isolation
GA   Genetic algorithm
GD   Grid diagram
GDFCM   Geodesic distance-based fuzzy c-medoid clustering method
GDFCMSA   Geodesic distance-based fuzzy c-medoid clustering method with
          simulated annealing
HEN   Heat exchanger network
HLMN   Human liver metabolic network
HSCN   Human signalling cancer network

LCC	Longest control chain
LTI	Linear time-invariant
mCLASA	Modified clustering large applications based on simulated annealing
MDS	Minimum driver node set
MER	Maximum energy recovery or minimum energy requirements
MIMO	Multiple-input and multiple-output
MNA	Modified nodal analysis
ncRNA	Non-protein coding ribonucleic acid
NOCAD	Network-based observability and controllability analysis of dynamical systems
OSLP	Optimal sensor location problem
PEEC	Partial element equivalent circuit
PFD	Process flow diagram
PI	Process integration
PSO	Particle swarm optimisation
RI	Resilience index
SA	Simulated annealing
SCC	Strongly connected components
SHM	Structural health monitoring
SM	Streams and matches
SS	State-space approach
TAC	Total annualised cost
TCO	Target control problem with objectives-guided optimisation
TOG	Time-ordered graph
TWC	Total walk count

Chapter 1
Introduction

Abstract The application of network science in the field of dynamical systems has enabled a set of powerful tools that can be used for the analysis of dynamical properties of complex systems. In this chapter, the most important trends and applications of this novel approach are introduced.

Keywords Dynamical systems · Controllability · Observability · Maximum matching · Network science · Input configuration · Output configuration · Robustness

1.1 Network Science-Based Analysis of Complex Systems

Nowadays, unlike a few decades ago, network science-based controllability and observability analysis of complex systems is a familiar phrase to most scientists. However, a few decades ago this was not typical. The importance of the relations between the elements in interconnected systems is becoming a focus of research. The analysis of the complex representation of systems demands novel methods that can efficiently handle such degrees of complexity. It is not a coincidence that Stephen W. Hawking (1942–2018) stated in an interview with The Mercury News in San Jose, California in January 2000: 'I think the next century will be the century of complexity'. Network science may be a possible answer to this challenge as it provides a simple but great set of tools to represent and analyse the dynamics within the elements of the system. The success of the methodology is unequivocal in terms of the number of papers published in this field as represented in Fig. 1.1.

In spite of the remarkable work of Erdős and Rényi published in 1960 [13], the success can be rather attributed to the work of Barabási et al. [6] and Liu et al. [35]. While the former researchers laid the foundation with regard to the structural analysis of connected systems by applying mathematical graphs, the latter, besides utilising mathematical graphs, rather dealt with the dynamics of networks. It is interesting that the essence of the latter publication had already been introduced in 1974 [32]. Therefore, the field became suitable for analysing complex and dynamical systems.

To review this topic, the most cited and important papers which deal with the dynamics of complex systems were surveyed. To determine the main research areas, a network of the keywords provided by the authors was generated (Fig. 1.2).

© The Author(s), under exclusive license to Springer Nature Switzerland AG 2020   1
D. Leitold et al., *Network-Based Analysis of Dynamical Systems*,
SpringerBriefs in Computer Science,
https://doi.org/10.1007/978-3-030-36472-4_1

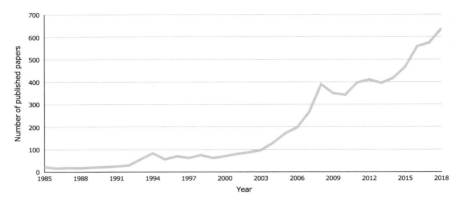

Fig. 1.1 The number of published articles in the field of network science-based controllability and observability analysis over recent years. The exponential growth of the number of articles is proof of the applicability of the approach in handling the challenges of the twenty-first century

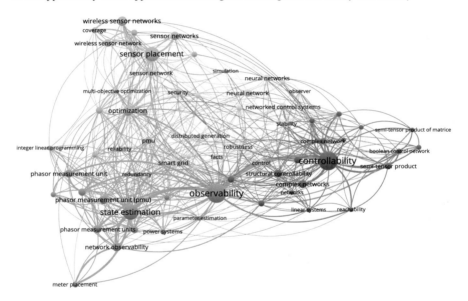

Fig. 1.2 The network of the most common keywords in the published articles in the field of network science-based controllability and observability analysis. The networked representation of the keywords highlights the most intensively researched topics and methods

As network science is a great tool to represent complex systems, some reviews have also been published. In [4], network science was utilised in the field of tourism and its benefits introduced through case studies known as Elba and Italy. A review concerning the applicability of network science in the fields of Geography and Regional Science has been published [12]. The advantages of network science in the study of complex biological systems have also been reviewed [18]. A more general review covers a range of disciplines, where the potential tools and concepts of network science were revealed [14]. Although these reviews are remarkable and deal with network science, they fail to deal with the methodology of the network science-based controllability and observability analysis of dynamical systems. Despite the enormous interest in such methodology, only two thorough reviews have been published in this field. The first was published 5 years after the publication of the fundamental article from 2011 [34]. This work discusses the main topics, namely the controllability of linear and nonlinear systems, observability analysis, trajectories and controllability of collective behaviour. The second one discussed the control principles of the methodology in the field of biology [30], namely in the determination of disease genes and drug targets, utilisation in neural and brain dynamic networks and interpretation of control energy. Although the first review is very detailed, it does not contain the latest results, and the second one is too specific to the field of biology. Thus, in the following, after a brief introduction to the methodology, a short summary of what progress has been made in this field since 2016 is presented.

1.2 Formalisation of the Network-Based Controllability and Observability Analysis

What is meant by a social network that is controllable? The answer is derived from the analysis of dynamical systems. Dynamical systems were created to provide assistance in our everyday lives, e.g. the steam engine, samovar and elevator. Probably the most important properties of these systems are their controllability and observability, which enable them to be handled and monitored. Network science applies the maximum matching algorithm to determine the inputs and outputs of a linear time-invariant (LTI) system to grant structural controllability and structural observability. An LTI system can be described according to its state-space representation that contains a state equation (Eq. 1.1) and output equation (Eq. 1.2). Although Eqs. 1.1 and 1.2 are represented as differential equations, the methodology can also be applied to discrete-time systems that are described by difference equations, the only requirement is the linearity of the system. Other classes of systems are not introduced here since they are beyond the scope of this work.

$$\dot{x} = \mathbf{A}x + \mathbf{B}u(+\mathbf{E}d) \tag{1.1}$$

$$y = \mathbf{C}x + \mathbf{D}u \tag{1.2}$$

In state-space representation, x stands for state variables, u represents the inputs, i.e. the actuators, d denotes the disturbances and y is the vector of outputs, i.e. the sensors of the system. The matrices $\mathbf{A}$, $\mathbf{B}$ and $\mathbf{E}$ define how state variables, inputs and disturbances influence changes to the state variables, while matrices $\mathbf{C}$ and $\mathbf{D}$ define how state variables and inputs influence the outputs, respectively. Assuming

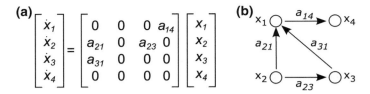

Fig. 1.3 The network representation of a state-transition matrix. **a** The state equation of the system with four state variables. **b** The directed network representation of the example LTI system, where the adjacency matrix of the network is defined by the state-transition matrix of the system (**A**)

that the number of internal state variables is N, and the number of inputs and outputs is M and K, respectively, then the sizes of the matrices are $\mathbf{A} \in \mathfrak{R}^{N \times N}$, $\mathbf{B} \in \mathfrak{R}^{N \times M}$, $\mathbf{E} \in \mathfrak{R}^{N \times N}$, $\mathbf{C} \in \mathfrak{R}^{K \times N}$ and $\mathbf{D} \in \mathfrak{R}^{K \times M}$. In this work, as in the majority of research, matrices D and E are defined as 0. If a deviation is made from this, the reader will be informed.

A dynamical system is said to be controllable, if it can be derived from the initial state into any desired final state within a finite period of time [23]. Controllability can be proven by Kalman's rank criterion, namely if the rank of the controllability matrix, $\mathcal{C} = [\mathbf{B}, \mathbf{AB}, \dots, \mathbf{A}^{N-1}\mathbf{B}]$ is equal to the number of state variables ($rank(\mathcal{C}) = N$), then the system is structurally controllable. Observability is the mathematical dual of controllability. According to the definition of observability, the whole state ($\boldsymbol{x}$) of the system can be recovered based on the inputs $\boldsymbol{u}$ and the measurements of derivatives 1, $2, \dots, N-1$ of outputs $\boldsymbol{y}$. Observability can be proven by the observability matrix, $\mathcal{O} = [\mathbf{C}^T, (\mathbf{CA})^T, \dots, (\mathbf{CA}^{N-1})^T]^T$, as if its rank is equal to the number of state variables ($rank(\mathcal{O}) = N$), then the system is structurally observable.

To ensure controllability (or observability) using a minimum number of inputs (or outputs), a brute force approach should generate $2^N - 1$ configurations of matrix $\mathbf{B}$ (or $\mathbf{C}$). To solve this challenging task, the maximum matching algorithm was utilised on the network representation of matrix $\mathbf{A}$ to select the driver [35] and sensor [37] nodes which ensure the controllability and observability of the system.

Based on the structure matrix of the state-transition matrix $\mathbf{A}$, a directed network can be created, where an edge from node i to j is present if state variable x_j appears on the right-hand side of the state equation of state variable x_i. A simple example can be seen in Fig. 1.3.

The *maximum matching algorithm* determines the maximum disjoint edge set for an undirected, bipartite graph. Two edges are disjointed if they have no common endpoint. To apply the maximum matching algorithm to a directed network, the network needs to be converted into an undirected, bipartite graph. For this purpose, two nodes need to be assigned to in the bipartite representation for each node from the directed network. Then, the first set of nodes in the bipartite representation is the *out*, while the second is the *in* representation of the nodes in the directed version. Then, the maximum disjoint edge set is determined. The resultant matched nodes are the endpoints of the resultant disjoint edges. In the directed graph representation, a node

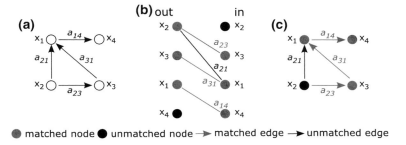

Fig. 1.4 Maximum matching in directed networks. **a** Directed network representation of an LTI system. **b** Undirected, bipartite graph of the directed network. The result of maximum matching is noted as follows: matched nodes and edges are denoted by the colour red, while unmatched ones by black. **c** The result of maximum matching in the directed network

is matched if its bipartite version in the *in* set is matched, otherwise it is unmatched. A simple example of the conversion and result of the maximum matching can be seen in Fig. 1.4.

The resultant unmatched nodes in the directed network representation are the sensor nodes, i.e. the nodes, where sensors should be placed to grant structural observability of the system. Since controllability is the mathematical dual of observability, the unmatched nodes generated for the network based on the transposition of the state-transition matrix are the driver nodes, where actuators should be placed to ensure structural controllability of the system. This methodology only provides structural analysis of a system, i.e. it exclusively deals with the structural architecture and ignores edge weights. The determined driver and sensor nodes are also just structural positions in the system, the methodology does not assign any parameters to matrices **B** and **C** or vectors u and y.

It is important to mention that the result of maximum matching is not unique and for the same network more solutions with the same number of matched nodes can exist. This results in different input and output configurations that result in different behaviours of complex systems. Nevertheless, an exceptional case is known: with perfect matching the number of unmatched nodes is zero. In this case, one driver node grants controllability independently of its location. To understand the control mechanism, Liu et al. introduced the concept of stem and cycle [36]. A stem is a directed path in the network that starts with a driver node. Cycles are controlled by the inputs, which possess at least one node that has an incoming edge from a controlled node. A node is controlled, if it is a member of the stem or a controlled cycle. Unfortunately, perfect matching can occur for parts of the network as well, namely Hamiltonian cycles are matched by themselves, therefore, in some unique situations, control of the unmatched nodes is insufficient to grant controllability. Two approaches handle this problem, referred to as the signal sharing method that addresses the problem concerning the strongly connected components (SCCs) [37], and the path-finding method that identifies another appropriate maximum matching, or assigns additional driver (or sensor) nodes to the Hamiltonian cycles. The mechanism of both methods is introduced in Appendix A.1.

1.3 Recent Trends in Network-Based Dynamical System Analysis

Based on Fig. 1.2 and following the aforementioned train of thought, some fields that have been researched intensively were determined and introduced one by one. Our review contains papers from 2016 onward as a very thorough review was published in 2016 about the methodology [34].

Firstly, the articles that propose methods to determine the driver and sensor nodes in a different way are introduced, i.e. in these papers, the input and output configurations were determined by an algorithm that did not utilise the maximum matching algorithm but was inspired by it and its efficiency. Usually, in these papers, the input and output configurations were optimised according to an objective function and, in the context addressed, the proposed method provided a better solution. The approaches can be seen in Table 1.1.

The methodology is defined in such a general way, that it can be applied to any topology, independent of its origin. Maybe this is the essence of the methodology and its success. However, the resultant configurations can result in different behaviours as previously mentioned. In Table 1.2, papers are presented that deal with the extension of the methodology according to additional constraints and objectives.

The provided input and output configurations are designed using a minimum number of driver and sensor nodes. This can result in an exceptionally high or even physically impossible energy demand, high relative degree or unfavourable dynamic behaviours. Thus, numerous papers have dealt with the improvement of the original methodology. Qualitative measures and heuristic approaches have been utilised to further refine the designed configurations as can be seen in Table 1.3.

Since for highly interconnected networks only one actuator can be sufficient, the robustness of the proposed solutions is questionable. Thus, some papers have analysed the robustness of the controllability or observability by excluding nodes or edges from the topology. Others have analysed the robustness using cascade models. A summary of these articles can be seen in Table 1.4.

The interest in network science-based system analysis has also spread to applications, not just to analytical improvements. The methodology has already been applied to several fields such as in biology, heat exchanger networks as well as quantum networks. The methodology is utilised with the greatest degree of success in the field of biology [65]. The analysed networks include the protein–protein interaction (PPI) networks, the neuronal network of Caenorhabditis elegans, neurochemical network of a rat brain, Saccharomyces cerevisiae cell cycle networks, epithelial–mesenchymal transition (EMT) regulatory networks, the regulatory network of myeloid differentiation and Th differentiation network. Another application is to realise pinning control

Table 1.1 Papers that propose alternative methods to determine input and output configurations

Description	Year	Refs.
A controllability Gramian-based metric was introduced and used to determine the globally optimal or near-optimal placement of actuators and sensors according to a greedy algorithm	2016	[50]
The importance of inaccessible strongly connected components (ISCC) was introduced in this work, and a branch-and-bound technique proposed to generate the minimum controlled vertex set	2016	[70]
A non-maximum matching-based (even though it was mentioned) method was proposed to study the controllability of networks composed of random tree structures according to the controllability index	2016	[7]
The controllability index was introduced to determine the relationship between system controllability and the controllability matrix	2017	[5]
The controllability of flow-conservation networks was also analysed without applying the maximum matching-based methodology	2017	[73]
Driver nodes were determined based on a dominating set. Hierarchical control and community dynamics were also analysed	2017	[51]
The Lyapunov stability theory-based estimation of states was conducted by a kind of output-feedback pinning observer	2017	[62]
A complete selection rule set was generated by a general procedure to grant structural controllability	2017	[63]
Distributed local game matching was utilised to grant structural controllability. With the minimisation of the longest control path, the cost of control was significantly reduced	2018	[28]
The structural, dynamical and symbolic observability of linear as well as nonlinear dynamical systems and dynamical networks was determined according to several methods	2018	[1]
A new network representation referred to as q-snapback was introduced, and the network centralities were interpreted. In the new representation, controllability was defined and robustness was examined against the removal of nodes and edges. The analysis showed that the degree of robustness was stronger than in other representations	2018	[38]
Non-matching-based structural controllability analysis of large interconnected systems was introduced	2018	[53]
A non-matching-based solution to observability in the case of nonlinear systems was provided, that is faster than existing methods	2018	[27]

that is concerned with the aforementioned robustness problem of the methodology in highly interconnected systems. The articles that introduce the applicability of the methodology are summarised in Table 1.5.

Table 1.2 Papers that extended the methodology with constraints and suggestions

Description	Year	Refs.
Higher order MIMO LTI systems were analysed as multi-layer networks	2016	[59]
Network motifs were revealed and the authors pointed out that structure-based analysis is unsuitable to characterise controllability in Boolean Networks if the dynamics are given and must be taken into account	2016	[17]
Based on network centrality measures, the probability of a node being a driver node was quantified. Driver nodes were classified based on how critical they are in terms of control	2016	[29]
The controllability of temporal networks was analysed by control centrality as a sequence of characteristic subgraphs and as a time-ordered graph (TOG)	2016	[31]
A minimum steering node set was introduced to determine the inaccessible set of nodes that must be controlled	2016	[66]
The structural properties of the control graphs of LTI systems were analysed	2016	[2]
The paper proposed a methodology to identify different maximum matchings of complex networks	2016	[68]
Networks were augmented while the structural controllability was conserved. Nodes were also classified based on how a newly attached node changes the number of driver nodes	2016	[55]
Controllability was analysed based on the degree correlation and it was highlighted that the number of driver nodes is strongly dependent on the degree correlation in both directed and undirected networks	2016	[42]
The paper called attention to the types of connections between state variables. These types are typical in real dynamical systems but not in other real networks that are not dynamical systems, e.g. social networks	2017	[24]
This work highlighted that structural controllability does not imply physical controllability, namely the energy cost can be physically impossible for the minimised driver nodes proposed by maximum matching	2017	[56]
Strong structural controllability was evaluated and edge controllability was expanded to all complex networks. The controllability of an arbitrary node was determined by its local weighted structure	2017	[44]
Canonical forms for controlling complex networks were proposed, where the minimum number of driver nodes was equal to one	2017	[58]
The maximum matching-based minimum number of driver nodes in systems with multidimensional node dynamics was analysed	2017	[64]
Simulated annealing was utilised in rewiring undirected networks to analyse the effect of degree correlation on controllability. The results showed that in undirected cases the effect complies with special rules and is not as clear as in the case of directed networks	2017	[40]
A framework was proposed to determine driver nodes for pinning control. The framework was proposed to assist in the diagnosis of dementia	2018	[52]
Structural and strong structural controllability of certain families of undirected networks via combinatorial constructions was analysed	2018	[41]
A method was proposed to determine the minimum number of edges that need to be added to the system in order to grant structural controllability (**A** and **B** are given)	2018	[10]

Table 1.3 Papers that improved the methodology in order to reduce costs, save energy or limit unfavourable dynamics

Description	Year	Refs.
Based on the physical theory, the longest control chain (LCC) was proposed and an LCC-based strategy was suggested to reduce the control energy drastically that diverges if the structural controllability is granted by the minimum number of driver nodes	2016	[11]
The efficacy of the control (or control centrality of a single driver) was determined in this paper and a method to identify controllable nodes also proposed	2016	[16]
The control range was interpreted to quantify the capacity of driver nodes in maintaining the system through chains and rings	2016	[15]
A method was suggested to determine complete schemes for eliminating redundant driver nodes such that the system remains structurally controllable	2017	[22]
The paper provided a centrality measures-based and a set covering-based methods to assign additional driver nodes to the system in order to decrease the relative degree of the system	2018	[25]
The paper provided two simulated annealing-based methods to assign additional driver nodes to the system in order to decrease the relative degree of the system	2018	[26]
Several constructive algorithms to minimise the control energy were proposed and a heuristic principle was suggested to reduce the controlling cost	2018	[33]
Two methods to determine minimal-cost actuator placement were proposed to guarantee symmetric structural controllability	2018	[48]

1.4   Motivation and Outline of the Book

Based on the previously presented overview, it is clearly visible that the network-based analysis of dynamical systems became widely used.

This book consists of four aims:

- to introduce connection types that are typical in real systems and networks, moreover, is very important in terms of the proper utilisation of the methodology,
- to present control and sensor placement approaches that can improve the dynamical properties of the designed system,
- to suggest a workflow for the systematic network science-based analysis of dynamical systems,
- and to introduce an Octave- and MATLAB-compatible toolbox referred to as NOCAD (Network-based Observability and Controllability Analysis of Dynamical Systems) to support the analysis with an open-source platform-independent tool.

Chapter 2 **(Effect of connection types)** introduces the four motifs that are typical in real dynamical systems and methods for maximum matching-based driver and sensor node selection. The commonly used benchmark examples are evaluated against

Table 1.4 Papers that dealt with robustness and proposed a method to reduce vulnerability

Description	Year	Refs.
The control vulnerability of Erdős–Rényi (ER), small world (SW) and scale-free (BA) networks were analysed by degree and betweenness-based attack strategies. The resultant weak points of the networks are of the maximum degree in ER networks of the maximum betweenness in SW networks, and hubs in BA networks. It was also presented that real networks respond in a significantly different manner with regard to attacks than artificially generated ones	2016	[39]
The robustness of controllability against edge removal was analysed	2016	[57]
The control cost was defined in order to evaluate the robustness of networks against cascading failures. The results show that the allocation of different edge capacities and control inputs can strengthen the robustness of the system	2017	[9]
This paper highlighted that the feedback structure plays a crucial role in selecting suitable robust control strategies	2017	[71]
Nodes and drivers were classified and the robustness of their roles against rewiring as well as the addition of nodes and edges analysed in the gene network that controls the flower development of Arabidopsis thaliana	2018	[60]
The robustness of edge controllability was analysed according to the strength of interactions between state variables. The edges were also categorised based on their role in terms of control. As a result, the dominant factors concerning robustness were degree distribution and interaction strength in directed and undirected networks, respectively	2018	[43]
The functional robustness of controllability of the power grid in China was analysed against cascading failures by network measures like the vulnerability index, degree, degree distribution, average path length, clustering coefficient and network diameter	2018	[61]
The effect of a cascading failure by the controllability of node attacking was analysed in interdependent complex networks	2018	[72]

real dynamical systems and conclusions are drawn about the proper utilisation of the methodology.

Chapter 3 (**Reduction of relative degree by additional drivers and sensors**) introduces five methods to design input and output configurations such that an initially defined relative degree is not exceeded. The proposed methods utilise closeness and betweenness centrality measures, the set covering method, simulated annealing, and fuzzy c-medoid approaches. The performance of each method is examined according to four real examples.

Chapter 4 (**Structural analysis of dynamical complexity of HENs**) systematically analyses more than 600 Heat Exchanger Networks. This huge amount of networks is derived from almost 50 heat integration problems solved by more than 20 different methods. Each HEN is then converted into three different network representations that are analysed by in excess of 50 network-based measures in order to determine structural features that can predict dynamical characteristics. The typical

Table 1.5 Papers that present the applicability of the methodology

Description	Year	Refs.
Therapeutic implications of driver nodes in disease control based on Human Cancer Signalling Networks (HCSN) were examined and boundaries defined	2016	[46]
The concept of steering node was introduced and then this kind of nodes were analysed in biomolecular networks like the metabolic network of the liver	2017	[67]
The methodology was utilised in terms of the pinning control of both SISO and MIMO systems that can be non-identical, nonlinear and linear. Linear time-varying systems were also discussed	2017	[8]
Driver nodes in biological networks such as HCSNs were determined as the best drug targets in a system that has malfunctioned and could potentially cause disorders	2017	[47]
A model of synaptic plasticity was proposed and utilised in the neuronal wiring map of C.elegans to determine optimum long-distance synaptic connections that were addressed as a key feature of control	2017	[3]
The controllability of a temporal and static functional brain was evaluated by control centrality measures	2017	[69]
The degree of controllability based on the controllability Gramian was evaluated in terms of drinking events	2017	[54]
The controllability of heterogeneous interdependent group systems for both directed and undirected cases was investigated	2018	[45]
From the perspective of network controllability, personalised driver mutation profiles were identified and evaluated based on improved precision, effectiveness and risk assessment	2018	[19]
Quantum networks were analysed and were all structurally controllable by a single driving signal	2019	[49]
The methodology has been a great success in the field of biology. This review article presents the applications of the methodology that have already been implemented in different biological networks	2019	[65]
Personalised network control model (PNC) was proposed to identify the personalised driver genes	2019	[20]
For combination therapy, OptiCon methods was proposed to identify synergistic regulators as candidates	2019	[21]

motifs, which are at the same time the elementary building blocks, that appear in Heat Exchanger Networks are highlighted and a community analysis is also presented.

Chapter 5 (The NOCAD toolbox) introduces the aforementioned toolbox that is compatible with the open-source Octave and MATLAB for industrial use. The toolbox enables the analysis expounded on in previous chapters to be conducted, and facilitates the analysis of any network topology. The implemented methods are introduced through an artificial example of a network that highlights the differences between the measures generated.

References

1. Aguirre, L.A., Portes, L.L., Letellier, C.: Structural, dynamical and symbolic observability: from dynamical systems to networks. PloS One **13**(10), e0206180 (2018)
2. Alwasel, B., Wolthusen, S.D.: Recovering structural controllability on erdős-rényi graphs via partial control structure re-use. In: International Conference on Critical Information Infrastructures Security, pp. 293–307. Springer (2014)
3. Badhwar, R., Bagler, G.: A distance constrained synaptic plasticity model of C. elegans neuronal network. Phys. A: Stat. Mech. Its Appl. **469**, 313–322 (2017)
4. Baggio, R., Scott, N., Cooper, C.: Network science: A review focused on tourism. Ann. Tour. Res. **37**(3), 802–827 (2010)
5. Bai, Y.-N., Wang, L., Chen, M.Z.Q., Huang, N.: Controllability emerging from conditional path reachability in complex networks. Int. J. Robust Nonlinear Control **27**(18), 4919–4930 (2017)
6. Barabási, A.-L., Albert, R.: Emergence of scaling in random networks. Science **286**(5439), 509–512 (1999)
7. Caro-Ruiz, C., Pavas, A., Mojica-Nava, E.: Controllability criterion for random tree networks with application to power systems. In: 2016 IEEE Conference on Control Applications (CCA), pp. 137–142. IEEE (2016)
8. Chen, G.: Pinning control and controllability of complex dynamical networks. Int. J. Autom. Comput. **14**(1), 1–9 (2017)
9. Chen, S.-M., Xu, Y.-F., Nie, S.: Robustness of network controllability in cascading failure. Phys. A: Stat. Mech. Its Appl. **471**, 536–539 (2017)
10. Chen, X., Pequito, S., Pappas, G.J., Preciado, V.M.: Minimal edge addition for network controllability. IEEE Trans. Control Netw. Syst. (2018)
11. Chen, Y.-Z., Wang, L.-Z., Wang, W.-X., Lai, Y.-C.: Energy scaling and reduction in controlling complex networks. R. Soc. Open Sci. **3**(4), 160064 (2016)
12. Ducruet, C., Beauguitte, L.: Spatial science and network science: review and outcomes of a complex relationship. Netw. Spat. Econ. **14**(3–4), 297–316 (2014)
13. Erdős, P., Rényi, A.: On the evolution of random graphs. Publ. Math. Inst. Hung. Acad. Sci. **5**(1), 17–60 (1960)
14. Estrada, E., Fox, M., Higham, D.J., Oppo, G.-L.: Network science: complexity in nature and technology. Springer Science & Business Media (2010)
15. Gao, X.-D., Shen, Z., Wang, W.-X.: Emergence of complexity in controlling simple regular networks. EPL (Eur. Lett.) **114**(6), 68002 (2016)
16. Gao, X.-D., Wang, W.-X., Lai, Y.-C.: Control efficacy of complex networks. Sci. Rep. **6**, 28037 (2016)
17. Gates, A.J., Rocha, L.M.: Control of complex networks requires both structure and dynamics. Sci. Rep. **6**, 24456 (2016)
18. Gosak, M., Marković, R., Dolenšek, J., Rupnik, M.S., Marhl, M., Stožer, A., Perc, M.: Network science of biological systems at different scales: a review. Phys. Life Rev. **24**, 118–135 (2018)
19. Guo, W.-F., Zhang, S.-W., Liu, L.-L., Liu, F., Shi, Q.-Q., Zhang, L., Tang, Y., Zeng, T., Chen, L.: Discovering personalized driver mutation profiles of single samples in cancer by network control strategy. Bioinformatics **34**(11), 1893–1903 (2018)
20. Guo, W.-F., Zhang, S.-W., Zeng, T., Li, Y., Gao, J., Chen, L.: A novel network control model for identifying personalized driver genes in cancer. bioRxiv, 503565 (2019)
21. Hu, Y., Chen, C.-H., Ding, Y.-Y., Wen, X., Wang, B., Gao, L., Tan, K.: Optimal control nodes in disease-perturbed networks as targets for combination therapy. Nat. Commun. **10**(1), 2180 (2019)
22. Huang, W., Xi, Y., Xu, Y., Gan, Z.: Eliminating redundant driver nodes with structural controllability guarantee. In: 2017 36th Chinese Control Conference (CCC), pp. 347–352. IEEE (2017)
23. Kalman, R.E.: Mathematical description of linear dynamical systems. J. Soc. Ind. Appl. Math., Ser. A: Control. **1**(2), 152–192 (1963)

24. Leitold, D., Vathy-Fogarassy, Á., Abonyi, J.: Controllability and observability in complex networks-the effect of connection types. Sci. Rep. **7**(1), 151 (2017)
25. Leitold, D., Vathy-Fogarassy, A., Abonyi, J.: Design-oriented structural controllability and observability analysis of heat exchanger networks. Chem. Eng. Trans. **70**, 595–600 (2018)
26. Leitold, D., Vathy-Fogarassy, A., Abonyi, J.: Network distance-based simulated annealing and fuzzy clustering for sensor placement ensuring observability and minimal relative degree. Sensors **18**(9), 3096 (2018)
27. Letellier, C., Sendiña-Nadal, I., Bianco-Martinez, E., Baptista, M.S.: A symbolic network-based nonlinear theory for dynamical systems observability. Sci. Rep. **8**(1), 3785 (2018)
28. Li, G., Deng, L., Xiao, G., Tang, P., Wen, C., Hu, W., Pei, J., Shi, L., Stanley, H.E.: Enabling controlling complex networks with local topological information. Sci. Rep. **8**(1), 4593 (2018)
29. Li, J., Dueñas-Osorio, L., Chen, C., Berryhill, B., Yazdani, A.: Characterizing the topological and controllability features of us power transmission networks. Phys. A: Stat. Mech. Its Appl. **453**, 84–98 (2016)
30. Li, M., Gao, H., Wang, J., Wu, F.-X.: Control principles for complex biological networksLi et al. Control principles for biological networks. Brief. Bioinform. (2018)
31. Li, X., Yao, P., Pan, Y.: Towards structural controllability of temporal complex networks. Complex Systems and Networks, pp. 341–371. Springer, Berlin (2016)
32. Lin, C.-T.: Structural controllability. IEEE Trans. Autom. Control **19**(3), 201–208 (1974)
33. Lindmark, G., Altafini, C.: Minimum energy control for complex networks. Sci. Rep. **8**(1), 3188 (2018)
34. Liu, Y.-Y., Barabási, A.-L.: Control principles of complex systems. Rev. Mod. Phys. **88**(3), 035006 (2016)
35. Liu, Y.-Y., Slotine, J.-J., Barabási, A.-L.: Controllability of complex networks. Nature **473**(7346), 167 (2011)
36. Liu, Yang-Yu., Slotine, Jean-Jacques, Barabási, Albert-László: Control centrality and hierarchical structure in complex networks. Plos one **7**(9), e44459 (2012)
37. Liu, Y.-Y., Slotine, J.-J., Barabási, A.-L.: Observability of complex systems. Proc. Natl. Acad. Sci. **110**(7), 2460–2465 (2013)
38. Lou, Y., Wang, L., Chen, G.: Toward stronger robustness of network controllability: a snapback network model. IEEE Trans. Circuits Syst. I: Regul. Pap. **65**(9), 2983–2991 (2018)
39. Lu, Z.-M., Li, X.-F.: Attack vulnerability of network controllability. PloS One **11**(9), e0162289 (2016)
40. Ming, X., Chuan-Yun, X., Ke-Fei, C.: Effect of degree correlations on controllability of undirected networks. Acta Phys. Sin. **66**(2) (2017)
41. Mousavi, S.S., Haeri, M., Mesbahi, M.: On the structural and strong structural controllability of undirected networks. IEEE Trans. Autom. Control **63**(7), 2234–2241 (2018)
42. Nie, S., Wang, X.-W., Wang, B.-H., Jiang, L.-L.: Effect of correlations on controllability transition in network control. Sci. Rep. **6**, 23952 (2016)
43. Pang, S.-P., Hao, F.: Effect of interaction strength on robustness of controlling edge dynamics in complex networks. Phys. A: Stat. Mech. Its Appl. **497**, 246–257 (2018)
44. Pang, S.-P., Wang, W.-X., Hao, F., Lai, Y.-C.: Universal framework for edge controllability of complex networks. Sci. Rep. **7**(1), 4224 (2017)
45. Pei, H.-Q., Chen, S.-M.: Controllability of heterogeneous interdependent group systems under undirected and directed topology. Chin. Phys. B **27**(10), 108901 (2018)
46. Ravindran, V., Sunitha, V., Bagler, G.: Controllability of human cancer signaling network. In: 2016 International Conference on Signal Processing and Communication (ICSC), pp. 363–367. IEEE (2016)
47. Ravindran, V., Sunitha, V., Bagler, G.: Identification of critical regulatory genes in cancer signaling network using controllability analysis. Phys. A: Stat. Mech. Its Appl. **474**, 134–143 (2017)
48. Romero, O., Pequito, S.: Actuator placement for symmetric structural controllability with heterogeneous costs. IEEE Control Syst. Lett. **2**(4), 821–826 (2018)

49. Siomau, M.: Any quantum network is structurally controllable by a single driving signal. Quantum Inf. Process. **18**(1), 1 (2019)
50. Summers, T.H., Cortesi, F.L., Lygeros, J.: On submodularity and controllability in complex dynamical networks. IEEE Trans. Control Netw. Syst. **3**(1), 91–101 (2016)
51. Sun, P.G., Ma, X.: Dominating communities for hierarchical control of complex networks. Inf. Sci. **414**, 247–259 (2017)
52. Tahmassebi, A., Amani, A.M., Pinker-Domenig, K., Meyer-Baese, A.: Determining disease evolution driver nodes in dementia networks. In: Medical Imaging 2018: Biomedical Applications in Molecular, Structural, and Functional Imaging. International Society for Optics and Photonics, vol. 10578, p. 1057829 (2018)
53. Van Der Woude, J., Boukhobza, T., Commault, C.: On structural behavioural controllability of linear discrete time systems with delays. Syst. Control Lett. **119**, 31–38 (2018)
54. Villasanti, H.G., Passino, K.M., Clapp, J.D., Madden, D.R.: A control-theoretic assessment of interventions during drinking events. IEEE Trans. Cybern. (2017)
55. Wang, J., Yu, X., Stone, L.: Effective augmentation of complex networks. Sci. Rep. **6**, 25627 (2016)
56. Wang, L.-Z., Chen, Y.-Z., Wang, W.-X., Lai, Y.-C.: Physical controllability of complex networks. Sci. Rep. **7**, 40198 (2017)
57. Wang, L., Bai, Y.-N., Chen, M.Z.Q.: Structural controllability analysis of complex networks. In: 2016 35th Chinese Control Conference (CCC), pp. 1225–1229. IEEE (2016)
58. Wang, L., Wang, L., Kong, Z.: Two controllable canonical forms for single input complex network. In: 2017 29th Chinese Control And Decision Conference (CCDC), pp. 1467–1472. IEEE (2017)
59. Wang, L., Chen, G., Wang, X., Tang, W.K.S.: Controllability of networked mimo systems. Automatica **69**, 405–409 (2016)
60. Wang P., Wang D., Lu, J.: Controllability analysis of a gene network for arabidopsis thaliana reveals characteristics of functional gene families. IEEE/ACM Trans. Comput. Biol. Bioinform. (2018)
61. Wang S., Zhang J., Yue X.: Multiple robustness assessment method for understanding structural and functional characteristics of the power network. Phys. A: Stat. Mech. Its Appl. **510**, 261–270 (2018)
62. Wang, W., Wan, Y., Liang, X.: State estimation for complex network with one step induced delay based on structural controllability and pinning control. In: Intelligent Computing, Networked Control, and Their Engineering Applications, pp. 575–584. Springer (2017)
63. Wang, X., Xi, Y., Huang, W., Jia, S.: Deducing complete selection rule set for driver nodes to guarantee network's structural controllability. IEEE/CAA J. Autom. Sin. (2017)
64. Wang, X.-W., Jiang, G.-P., Wu, X.: Structural controllability of complex dynamical networks with nodes being multidimensional dynamics. In: American Control Conference (ACC), pp. 5013–5019. IEEE (2017)
65. Wu, L., Li, M., Wang, J.-X., Wu, F.-X.: Controllability and its applications to biological networks. J. Comput. Sci. Technol. **34**(1), 16–34 (2019)
66. Wu, L., Li, M., Wang, J., Wu, F.-X.: Minimum steering node set of complex networks and its applications to biomolecular networks. IET Syst. Biol. **10**(3), 116–123 (2016)
67. Wu, L., Tang, L., Li, M., Wang, J., Wu, F.-X.: The MSS of complex networks with centrality based preference and its application to biomolecular networks. In: 2016 IEEE International Conference on Bioinformatics and Biomedicine (BIBM), pp. 229–234. IEEE (2016)
68. Yang, Y., Xie, G.: Mining maximum matchings of controllability of directed networks based on in-degree priority. In: 2016 35th Chinese Control Conference (CCC), pp. 1263–1267. IEEE (2016)
69. Yao, P., Li, C., Li, X.: The functional regions in structural controllability of human functional brain networks. In: 2017 IEEE International Conference on Systems, Man, and Cybernetics (SMC), pp. 1603–1608. IEEE (2017)
70. Yin, H., Zhang, S.: Minimum structural controllability problems of complex networks. Phys. A: Stat. Mech. Its Appl. **443**, 467–476 (2016)

71. Zañudo, J.G.T., Yang, G., Albert, R.: Structure-based control of complex networks with non-linear dynamics. Proc. Natl. Acad. Sci. **114**(28), 7234–7239 (2017)
72. Zhang, Z., Yin, Y., Zhang, X., Liu, L.: Optimization of robustness of interdependent network controllability by redundant design. PloS One **13**(2), e0192874 (2018)
73. Zhao, C., Zeng, A., Jiang, R., Yuan, Z., Wang, W.-X.: Controllability of flow-conservation networks. Phys. Rev. E **96**(1), 012314 (2017)

Chapter 2
Structural Controllability and Observability Analysis in Complex Networks

Abstract This chapter introduces the reader to the network-based analysis of the controllability and observability of dynamical systems and draws attention to the importance of dynamics between the state variables.

Keywords Dynamical systems · Controllability · Observability · Connection types · Dynamic behaviour · Network theory

2.1 Critical Evaluation of the Methodology

Despite the groundbreaking successes of the network theory-based controllability and observability analysis, the methodology has also been received some critiques. Müller and Schuppert determined that in transcriptional networks the maximum matching-based method drastically overestimates the number of necessary inputs [9]. Sun, Cornelius, Kath and Motter also highlighted that the methodology needs further clarification because the method gives incorrect results for non-linear systems even for small examples [10]. This fact has also been evinced by Dunne, Williams and Martinez [5]. Another problem is that researchers examined the correlation between necessary inputs generated by the maximum matching algorithm, and structural properties, like degree distribution, but they did not take into account that the result of the maximum matching algorithm is not unique [13].

The most contestable point of the network-based analysis is that it is based on a static and structural view of the system. We wish to offer a solution to the previously mentioned problems by examining how system dynamics, and other network-related dynamics should be represented realistically. The usage of proper topology is important and a crucial part of network analysis, as this is the only way to emphasise dynamics in static representations. We introduce connection types according to the typical relationships of the state variables. To analyse how the determined connection types influence the controllability and observability of dynamical systems we developed an Octave- and MATLAB-compatible toolbox. We examined 35 example networks used in articles and found that 27 do not represent dynamical systems, or the dynamical interpretation of the network. By comparing them with 18 independently

D. Leitold et al., *Network-Based Analysis of Dynamical Systems*,
SpringerBriefs in Computer Science,
https://doi.org/10.1007/978-3-030-36472-4_2

selected dynamical systems, we revealed significant differences. While in dynamical systems the number of inputs and outputs does not change when the proper topology of the model is studied, in the case of other networks more than 95% of inputs and outputs disappeared because of the determined connection types. It is believed that this issue is very important as much research and numerous applications from the field of biology have utilised the methodology [8, 12].

In the following section, we present how the functional relationships of state variables define different types of connections, and propose a workflow to evaluate how these connection types influence the number of necessary actuators and sensors that ensure controllability and observability. As a result, it can be concluded that in most of the articles dealing with network science-based controllability and observability analysis, mainly the physical topologies of the systems are studied and their dynamics were not taken into account, while the analysis of the realistic state-transition matrix-based topologies can lead to significantly different conclusions.

This chapter is based on the previous work of the authors [7], so most of the figures and tables are taken from this open-source publication.

2.2   Connections Between the State Variables and Their Effect on Controllability and Observability

The connection between two state variables, i.e. nodes, is represented by the sub-matrix of $\mathbf{A}$ that belongs to nodes i and j:

$$\mathbf{A}^{(i,j)} = \begin{bmatrix} a_{ii} & a_{ij} \\ a_{ji} & a_{jj} \end{bmatrix} \tag{2.1}$$

When there is a connection between the x_i and x_j state variables, we can generate 2^3-1 different combinations of the non-zero elements (edges) (see Fig. 2.1). Diagonal elements represent loops that store information about the films an actor played in or describe a variable that has integrating characteristics, i.e. the value of a state variable depends on its previous value, that can be defined as

$$\frac{\mathrm{d}x_i}{\mathrm{d}t} = f(\boldsymbol{x}), \quad x_i \in \boldsymbol{x} \tag{2.2}$$

By interpreting the principles of conservation of mass, energy or momentum, it can be realised that several connections should contain loops and symmetrical edges. This dynamic can represent co-authoring in collaboration networks or friendship in social networks. In the case of a symmetrical relationship, the strength of the interaction (change in the state variable) is also a function of both variables, i.e. if the value of state variable x_i depends on x_j, then x_i has an effect on the value of x_j:

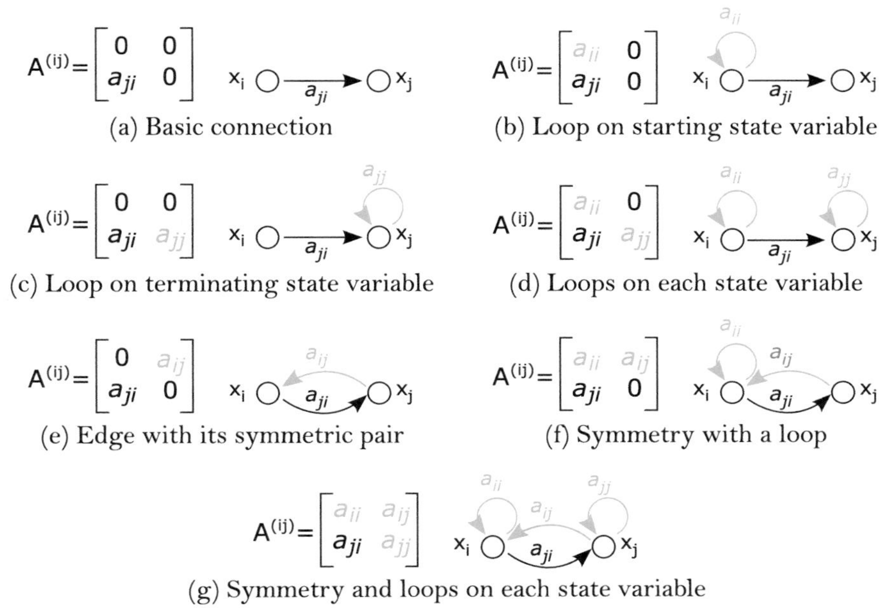

$$A^{(ij)} = \begin{bmatrix} 0 & 0 \\ a_{ji} & 0 \end{bmatrix} \qquad x_i \bigcirc \xrightarrow{a_{ji}} \bigcirc x_j$$

(a) Basic connection

$$A^{(ij)} = \begin{bmatrix} a_{ii} & 0 \\ a_{ji} & 0 \end{bmatrix} \qquad x_i \bigcirc \xrightarrow{a_{ji}} \bigcirc x_j$$

(b) Loop on starting state variable

$$A^{(ij)} = \begin{bmatrix} 0 & 0 \\ a_{ji} & a_{jj} \end{bmatrix} \qquad x_i \bigcirc \xrightarrow{a_{ji}} \bigcirc x_j$$

(c) Loop on terminating state variable

$$A^{(ij)} = \begin{bmatrix} a_{ii} & 0 \\ a_{ji} & a_{jj} \end{bmatrix} \qquad x_i \bigcirc \xrightarrow{a_{ji}} \bigcirc x_j$$

(d) Loops on each state variable

$$A^{(ij)} = \begin{bmatrix} 0 & a_{ij} \\ a_{ji} & 0 \end{bmatrix} \qquad x_i \bigcirc \underset{a_{ji}}{\overset{a_{ij}}{\rightleftarrows}} \bigcirc x_j$$

(e) Edge with its symmetric pair

$$A^{(ij)} = \begin{bmatrix} a_{ii} & a_{ij} \\ a_{ji} & 0 \end{bmatrix} \qquad x_i \bigcirc \underset{a_{ji}}{\overset{a_{ij}}{\rightleftarrows}} \bigcirc x_j$$

(f) Symmetry with a loop

$$A^{(ij)} = \begin{bmatrix} a_{ii} & a_{ij} \\ a_{ji} & a_{jj} \end{bmatrix} \qquad x_i \bigcirc \underset{a_{ji}}{\overset{a_{ij}}{\rightleftarrows}} \bigcirc x_j$$

(g) Symmetry and loops on each state variable

Fig. 2.1 Dynamics of two state variables. a Basic connection without additional edges represents the 'logical' or 'structural' relationship between the elements. **b** Source state variable influences itself as well, for example, it represents a variable that has a capacity which changes due to the connection. **c** The change of the terminating state variable also depend on its value, e.g. accumulation. **d** Combined dynamics of types (**b**) and (**c**). **e** The symmetric edge-pair shows that the influence is undirected, or the strength of the influence depends on the other state variable, as the signal flows in causal bond graphs [1]. **f** Combined dynamics of types (**b**) and (**e**). **g** Combined dynamics of types (**d**) and (**e**)

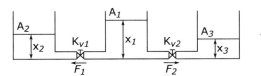

Fig. 2.2 Physical representation of water tanks. State variables represent water levels in the tanks. Flow rates F_1 and F_2 show how the water flows through the pipes from the first tank into the others

$$\frac{dx_i}{dt} = f(\boldsymbol{x}), \quad x_j \in \boldsymbol{x} \quad \wedge \quad \frac{dx_j}{dt} = f(\boldsymbol{x}'), \quad x_i \in \boldsymbol{x}'. \tag{2.3}$$

To illustrate connection types in a more tangible way, we illustrate them using a simple example. Let us consider three water tanks connected by two pipes (see Fig. 2.2.)

The difference between the physical topology and structure of the state-transition matrix (Eq. (2.4), where A_i is the area of tank i, K_{vi} represents the state of valve

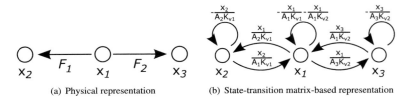

(a) Physical representation   (b) State-transition matrix-based representation

Fig. 2.3 Physical representation and state-transition matrix-based representation of water tanks. The difference between the physical structure and the structure of the properly modelled state-transition matrix is significant

i and x_i denotes the water level in tank i) can be seen in Fig. 2.3. It was supposed that the pressure/level is higher in tank 1 than in tank 2 or 3. Thus, the water can flow only in one direction, and the topology of the system remains unchanged. We use this supposition to ensure the linearity of the system. Although controllability of switching linear systems was examined, it is beyond the scope of this work [6].

$$\begin{bmatrix} \dot{x}_1 \\ \dot{x}_2 \\ \dot{x}_3 \end{bmatrix} = \begin{bmatrix} -\dfrac{1}{A_1 K_{v1}} - \dfrac{1}{A_1 K_{v2}} & \dfrac{1}{A_1 K_{v1}} & \dfrac{1}{A_1 K_{v2}} \\ \dfrac{1}{A_2 K_{v1}} & -\dfrac{1}{A_2 K_{v1}} & 0 \\ \dfrac{1}{A_3 K_{v2}} & 0 & -\dfrac{1}{A_3 K_{v2}} \end{bmatrix} \begin{bmatrix} x_1 \\ x_2 \\ x_3 \end{bmatrix} \qquad (2.4)$$

To analyse the effect of connection types, we defined four types of networks based on the four combinations of loops and edges. These four types are: networks that simply reflect the physical connections; networks with nodes having loops that represent the capacity or self-influencing of the state variables (the integrating behaviour of the tanks); networks with symmetric edge-pairs that reflect interactions between the state variables (mass balance); finally, networks with self-influence and interaction (Fig. 2.4). The first (Influence) and last (Self-influencing interaction) types were systematically introduced above through the example of water tanks. The second type (Self-influencing influence) can appear in the water tanks, e.g. if the first tank is placed higher than the other two, the water can flow from the first tank to the second and third tanks independently from the water level in these two tanks. The third type (interaction) can represent the dynamics of zero-order chemical reactions, radioactive decay or co-working in a collaboration network. Notwithstanding, the effect of intrinsic nodal dynamics on the number of driver nodes was examined [14], our approach differs, since it does not consider the different order of dynamics but the presence of first-order dynamics, therefore, the results also differ.

This example confirms that before generating input and output configurations the dynamical behaviour of the system must be examined more carefully, and the topology of the network has to be changed according to the required connection types. Figure 2.5 represents the flowchart of this suggested workflow.

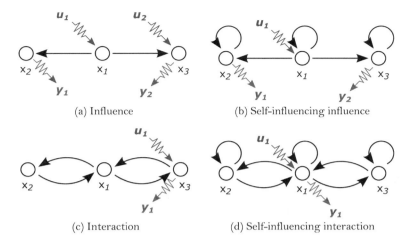

(a) Influence   (b) Self-influencing influence

(c) Interaction   (d) Self-influencing interaction

Fig. 2.4   The examined four types of water-tank network models with associated inputs (u) and outputs (y). a Physical connection between state variables. **b** Self-influencing represents integrating node dynamics. **c** Interaction represents balanced dynamics (material balance). **d** Detailed model of the system. Already this small example clearly shows how connection types change the number and location of the driver and sensor nodes. The driver and sensor nodes were determined by the path-finding method introduced in Sect. A.1 of the Appendix

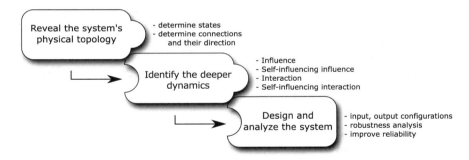

Fig. 2.5   Determined flowchart for system design and analysis. Since connection types can be determined according to the topology of the systems, we strongly recommend following the proposed workflow and taking into account deeper dynamics to obtain more accurate and reliable results

2.3   Benchmark Examples

In the literature, more or less the same set of networks is studied to demonstrate the effectiveness of network-based controllability and observability techniques. The main problem with these works is that majority of the studied problems do not represent dynamical processes. Considering tools of system theory we established dynamics that are typical in dynamical systems, and with this knowledge we exam-

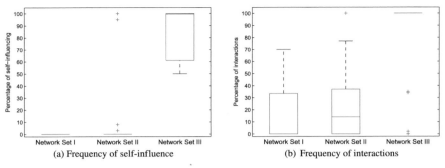

(a) Frequency of self-influence   (b) Frequency of interactions

Fig. 2.6 Initial percentage of self-influence and interactions in network sets. By examining real networks and dynamical systems, we get unequivocal results of structural differences. **a** Network Set I does not exhibit self-influence, and Network Set II does not exhibit them except in four cases, and only two of them exhibit more than 8% of self-influencing edges in their topologies. In contrast, more than 50% of self-influencing interactions in Network Set III are observed, i.e. more than half of the state variables influence themselves in dynamical systems. The median is almost 100%, so these dynamics are usual in dynamical systems. **b** In Network Set I usually less than 35% of interaction type edges are observed. In the case of Network Set II, results are a little higher. In Network Set III only four systems contain less than 100% of interactions, so in dynamical systems these dynamics is always present

ined how these dynamics appear, and how they influence the properties of networks. To answer these questions we grouped networks into three subgroups according to their dynamics. Firstly, *Network Set I* contains networks which are examined in articles, and can represent dynamical processes. Topologies in this set usually originate from the field of regulatory, transcriptional, neuronal, power grid or watershed networks. In these networks, dynamics are interpretable (in the sense of systems), i.e. some kind of capacity, or conservation law is observable between elements, such as Kirchhoff's circuit laws, even if it does not appear in representations. Secondly, *Network Set II* contains such networks, where dynamics do not appear, e.g. in a social network or in an email network the information or the message can be propagated without limitations. Furthermore, the Internet, citation networks and food webs belong to this set also. It is interesting why dynamical systems and their state-transition matrices are not included in articles. Therefore, in *Network Set III* we included state-transition matrices of real dynamical systems, thus we were sure that results of these networks provide the real behaviour of dynamical systems. Sizes and short descriptions of networks can be found in Appendix A.2. It should be mentioned that although in networks of *Network Set II* dynamical processes, e.g. information transfer [11] and transportation, knowledge diffusion [2–4] can take place. During the analysis of the networks, the connections are not interpreted from the viewpoints of the related dynamical processes.

Firstly, to prove that the established connection types are important parts of dynamical systems, and are not of other networks, we analysed how many self-influences and interactions are in these topologies (Fig. 2.6). Results clearly show that these connection types are fundamental parts of well-designed dynamical systems, but rare in real networks, where network processes were not interpreted. We concluded that this meaningful information about networks can be found in the literature, because only physical topologies were used to identify inputs and outputs instead of real state-space-based topology. Consequently, the determined number of driver and sensor nodes could be highly overestimated as is also shown by Müller and Schuppert [9].

2.4 Effect of the Connection Types

The extension of the networks with new edges due to self-influences and interactions changes the required number of the driver and sensor nodes. The results can be seen in Fig. 2.7. It is an interesting fact that networks from Network Set I reacted differently in terms of node dynamics, i.e. exhibited self-influence influence, compared to networks from Network Set II. In more detail, if self-influence appears in networks from Network Set I, then they show more willingness to reduce sensor nodes, while in the case of Network Set II the reduction is rather significant in driver nodes.

Nevertheless, the more important result is that networks from Network Set III did not show any changes with the newly added edges, i.e. we can assume that the determined node and edge dynamics are parts of dynamical systems. In contrast, other examples exhibit more than 95% of driver and sensor reduction if both dynamics were taken into account. This drastic difference shows the importance of the presence of node and edge dynamics in topologies. In addition, we note that networks from Network Sets I and II could provide wrong results in this field if the researched area was sensitive to the lack of dynamics. Detailed results can be found in Sect. A.3 of the Appendix.

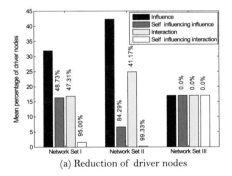
(a) Reduction of driver nodes

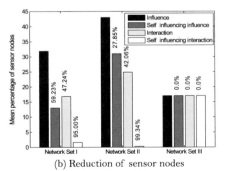
(b) Reduction of sensor nodes

Fig. 2.7 Proportion of driver and sensor nodes in networks with different connection types. Bar diagrams containing the mean of the proportion of driver (**a**) or sensor (**b**) nodes grouped according to network types. Apart from the influence, the numbers on the top of the bars show the reduction of driver and sensor nodes compared to the original

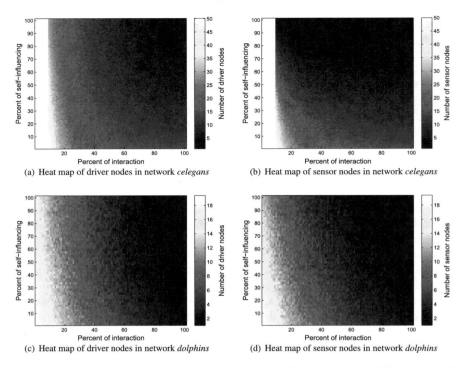

(a) Heat map of driver nodes in network *celegans*   (b) Heat map of sensor nodes in network *celegans*

(c) Heat map of driver nodes in network *dolphins*   (d) Heat map of sensor nodes in network *dolphins*

Fig. 2.8 **Number of driver and sensor nodes as a function of the presence of self-influence and interactions.** The heat map clearly shows that with the increase in self-influencing and interaction type edges the number of driver and sensor nodes decreased progressively. The white bars in heat maps of *C.elegans* network are caused by the initial existence of 9% of interaction-type connections

The reduction of driver and sensor nodes can be addressed in terms of the newly created strongly connected components (SCC). A self-influencing edge creates an SCC that contains one node, while an interaction creates an SCC that contains two nodes, respectively. It is equal to one-length and two-length cycles, that can be controlled by a stem easily (for details, see Sect. 1.2). This phenomenon clearly shows that the interpretation of self-influence on each node and the interpretation of interaction on each connected node-pair is not exaggerated: the number of driver or sensor nodes is progressively decreased by increasing the number of modified nodes and edges. To confirm our statement, we generated the heat map of network *C.elegans* from Network Set I and *dolphins* from Network Set II, that can be seen in Fig. 2.8. Furthermore, if both connection types appear in the network, then each connected component of a network can be controlled by one driver node and observed by one sensor node. Thus, with the help of Fig. 2.7 we can also conclude that networks from Network Set III contain more components than the others. As mentioned previously, the path-finding method was used to generate driver and sensor nodes. If we apply the signal sharing method, then the results are the same, only three cases differ slightly.

In terms of influence, i.e. the original network, for Network Set I the method assigned 0.17605% more sensor nodes, while for Network Set II 0.00356% more driver and 0.01046% more sensor nodes were determined compared to the path-finding method.

2.5   Discussion

Although the network science-based examination of the system controllability and observability is a very popular and fruitful methodology, it is still in its infancy. Besides positive results, some negative comments were also published. In this Chapter, we provided answers to criticisms using the determined connection types. With a novel workflow, how network-based analysis could generate more realistic results if the connections among the nodes and state variables would be defined based on more detailed analysis was illustrated. We also realised that the majority of the networks studied in the literature do not exhibit dynamical behaviour that could be interpreted as a (linear) state-space model. Network process-related dynamics should be interpreted in real networks, as in the examined topologies when the integrating behaviour and balance equation-related relations among the state variables were also taken into account, the number of necessary drivers and sensors decreased drastically compared to the analysis of oversimplified structural networks. Nevertheless, the reduced number of inputs and outputs results in a system with high relative degree that prevents fast and robust process monitoring and control.

References

1. Broenink, J.F.: Introduction to physical systems modelling with bond graphs. SiE whitebook on simulation methodologies, 31 (1999)
2. Chen, W., Lu, W., Zhang, N.: Time-critical influence maximization in social networks with time-delayed diffusion process. In: Twenty-Sixth AAAI Conference on Artificial Intelligence (2012)
3. Colizza, V., Barrat, A., Barthélemy, M., Vespignani, A.: The role of the airline transportation network in the prediction and predictability of global epidemics. Proc. Natl. Acad. Sci. **103**(7), 2015–2020 (2006)
4. Cowan, R., Jonard, N.: Network structure and the diffusion of knowledge. J. Econ. Dyn. Control **28**(8), 1557–1575 (2004)
5. Gates, A.J., Rocha, L.M.: Control of complex networks requires both structure and dynamics. Sci. Rep. **6**, 24456 (2016)
6. Klamka, J., Niezabitowski, M.: Controllability of switched linear dynamical systems. In: 2013 18th International Conference on Methods & Models in Automation & Robotics (MMAR), pp. 464–467. IEEE (2013)
7. Leitold, D., Vathy-Fogarassy, Á., Abonyi, J.: Controllability and observability in complex networks mohana the effect of connection types. Sci. Rep. **7**, 151 (2017)
8. Li, M., Gao, H., Wang, J., Wu, F-X.: Control principles for complex biological networksli et al. control principles for biological networks. Briefings Bioinf. (2018)
9. Müller, F.-J., Schuppert, A.: Few inputs can reprogram biological networks. Nature **478**(7369), E4 (2011)

10. Sun, J., Cornelius, S.P., Kath, W.L., Motter, A.E.: Comment on controllability of complex networks with nonlinear dynamics (2011). arXiv:1108.5739
11. Weng, L., Flammini, A., Vespignani, A., Menczer, F.: Competition among memes in a world with limited attention. Sci. Rep. **2**, 335 (2012)
12. Lin, W., Li, M., Wang, J.-X., Fang-Xiang, W.: Controllability and its applications to biological networks. J. Comput. Sci. Technol. **34**(1), 16–34 (2019)
13. Zhang, X., Lv, T., Yang, X.Y., Zhang, B.: Structural controllability of complex networks based on preferential matching. PloS One **9**(11), e112039 (2014)
14. Zhao, C., Wang, W.-X., Liu, Y.-Y., Slotine, J.-J.: Intrinsic dynamics induce global symmetry in network controllability. Sci. Rep. **5**, 8422 (2015)

Chapter 3
Reduction of Relative Degree by Optimal Control and Sensor Placement

Abstract However, the resulting proportions of driver and sensor nodes are particularly small when compared to the size of the system, and although structural controllability and observability is ensured, the system demands additional drivers and sensors to provide the small relative degree needed for fast and robust process monitoring and control. In this chapter, a centrality measures-based, two set covering-based and two clustering and simulated annealing-based methods are proposed to assign additional drivers and sensors to the dynamical systems.

Keywords Sensor placement · Network science · Fuzzy clustering · Simulated annealing · Structural observability · Relative degree · Network segmentation

3.1 Problem of Sensor Placement

The placement of sensors significantly affects the performance of identification, state estimation as well as fault detection and isolation (FDI) algorithms. The goal-oriented placement of sensors for dynamical systems is a challenging task [9]. Parameter estimation-oriented information entropy-based optimal sensor placement was investigated in [27]. Based on robust information entropy, a Bayesian sequential sensor placement algorithm for multi-type of sensor was also proposed [38]. Computational fluid dynamics (CFD) models were generated to predict wind-flow, that used the data of a sensor placement utilised with prediction-value joint entropy [28]. For fault detection and isolation, an incremental analytical redundancy relation (ARR)-based algorithm was introduced in [30]. Structural analysis-based sensor placement strategies place additional sensors to isolate otherwise undisguisable faults [12]. In the initial sensor-constrained problems, the number of sensors is given and the detection time has to be minimised [3], while as far as time-constrained problems are concerned, the aim is to minimise the number of required sensors that ensure a predefined detection time. In the case of dynamical systems, the previously mentioned detection time is related to the relative degree of the system [17]. Besides good FDI performance, the optimal placement of the sensors should also ensure the observability of the systems [5].

In the case of dynamical systems, if observability analysis is performed on a properly represented network, the problem concerning the observability analysis is reduced to a reachability problem [34]. In some cascaded systems, this highlights that the number of necessary sensors is significantly smaller than the size of the state variables, which makes the observation vulnerable and the relative degree, i.e. the minimum number of derivatives that is necessary to observe at least one signal of inputs, is very high. To deal with the vulnerability, Liu et al. proposed a methodology to grant robustness in the undirected representation of the system [22].

There is an almost infinite number of objectives that can be used to define a sensor placement problem [32]. In most of the cases, the optimisation problem is formulated as that the instrumentation cost should be minimised and the precision of reconciled values should be ensured in a reliable, resilient and robust manner [2]. As the problem is NP-hard, the mixed-integer programming-based approach is suitable for small-to-medium sized networks [35]. To handle the complexity of the problem, mostly a heuristic approach is followed. For example, to reconstruct the internal states of biochemical reaction systems with the minimum number of sensors a graphical method has been developed [23]. Among population-based optimisation algorithms, genetic algorithms (GAs) are the most frequently applied, e.g. GA was utilised to find a sensor configuration that minimises cost and maximises the reliability and observability of the system [6]. The main benefit of these gradient-free heuristic optimisation methods is that they can be utilised with a wide range of models [1]. Simulated annealing (SA) is also widely employed optimisation algorithm in sensor placement [8]. SA lends itself to be combined with other search algorithms, e.g. with local search heuristics [25] or other SA inner loop, which results in dual representation simulated annealing (DRSA) [33]. These techniques are often combined also in sensor placement, e.g. GA and SA were successfully integrated [37]. Recently, particle swarm optimisation (PSO) was applied in optimal sensor location problem (OSLP) [29] and in the design of an ultrasonic structural health monitoring (SHM) system [4]. According to these publications, heuristic optimisation techniques are efficient in finding a small number of sensors with less time-consuming computations [40].

In this chapter, to ensure observability and minimise the relative degree of dynamical systems, four methods of network segmentation are introduced. We propose one more method also, which although minimises relative degree cannot grant structural observability. From the presented methods one utilises closeness and betweenness centrality measures, two of them solve set covering problem, and two of them combine the meta-heuristic SA optimisation method with clustering. SA is chosen as it exhibits excellent levels of performance with regard to graph-represented combinatorial optimisation (CO). The use of simulated annealing is strengthened by the fact that the relative degree of the system can be interpreted as a path in a network, and simulated annealing models a random walk on the search graph. Based on this concept, SA was applied to place micro-hydropowers in a water supply network [31] and network alignment in biological systems [24].

As sensors practically group the observed state variables, sensor placement can be considered as a special k-medoid clustering problem. This concept has been utilised in [39], where fuzzy c-means clustering algorithm was utilised to characterise the spatial distribution of the sensor placement problem.

The clustering large applications (CLASA) algorithm [10] developed as a fast and robust solution to the well-known k-medoid clustering problem. The fundamental idea of the chapter is that instead of considering the distance between the cluster centres and the clustered objects, a goal-oriented sensor placement algorithm can be derived by the introduction of a problem-relevant objective function into the scheme of the CLASA algorithm. The second novelty of the chapter is that the random search of the resulted algorithm is fine-tuned by calculating the selection probabilities based on network distance-related fuzzy membership functions.

In the following, the theoretical foundations and nomenclature of the relative degree of multiple-input and multiple-output (MIMO) dynamical systems are introduced. Then, the proposed optimisation algorithms are introduced, and examples of application are provided. Finally, the results are concluded.

This chapter is based on the previous work of the authors [21], so most of the figures and tables are taken from this open-source publication.

3.2 Relative Degree-Based Cost Functions

To evaluate the 'physical closeness' or 'direct effect' of the control configuration of a nonlinear system with state equation $\dot{x} = f(x) + \sum_{j=1}^{m} g_j(x)u_j + \sum_{\kappa=1}^{p} w_\kappa(x)d_\kappa$ and output equation $y_i = h_i(x)$, relative degree is introduced as a structural measure of the initial sluggishness of the response [11]. The relative degree can be determined by the standard Lie derivative. Therefore, for scalar field $h_i(x)$ and vector field $f(x)$ it is defined as $\mathcal{L}_f h_i(x) = \sum_{l=1}^{n} (\partial h(x)/\partial x_l) f_l(x))$, where $f_l(x)$ denotes the row element l of $f(x)$. Higher order Lie derivatives are defined as $\mathcal{L}_f^k h_i(x) = \mathcal{L}_f \mathcal{L}_f^{k-1} h_i(x)$, while mixed Lie derivatives $\mathcal{L}_{gj} \mathcal{L}_f^{k-1} h_i(x)$ in an obvious way [11]. Then, relative degree r_i is defined for output y_i as the smallest integer for which $[\mathcal{L}_{g1} \mathcal{L}_f^{r_i-1} h_i(x) \cdots \mathcal{L}_{gm} \mathcal{L}_f^{r_i-1} h_i(x)] \neq [0 \cdots 0]$ with respect to input vector $\mathbf{u}$ if exists, otherwise $r_i = \infty$. The relative degree r_{ij} of output y_i with respect to input u_j is the smallest integer for which $\mathcal{L}_{gj} \mathcal{L}_f^{r_{ij}-1} h_i(x) \neq 0$ if exists, otherwise $r_{ij} = \infty$. The relationship between r_i and r_{ij} can be defined as $r_i = min(r_{i1}, r_{i2}, \ldots, r_{im})$. Analogously to r_{ij}, the relative degree $\rho_{i\kappa}$ of output y_i with respect to disturbance d_κ can be defined as the smallest integer $\mathcal{L}_{w_\kappa} \mathcal{L}_f^{\rho_{i\kappa}} h_i(x) \neq 0$ if exists, otherwise $\rho_{i\kappa} = \infty$.

For linear systems that are defined in Eqs. (1.1) and (1.2), r_{ij} can be defined as the smallest integer for which $c_i \mathbf{A}^{r_{ij}-1} b^j \neq 0$, where c_i is row i of $\mathbf{C}$ and b^j is column j of $\mathbf{B}$. The relative degree $\rho_{i\kappa}$ with respect to the disturbance d is the smallest integer for which $c_i \mathbf{A}^{\rho_{i\kappa}-1} e^\kappa \neq 0$, where e^κ denotes column κ of $\mathbf{E}$. By utilising the network representation of any linear or linearised system, the relative degree r_{ij} can be defined as the length of the shortest path ℓ_{ij} from sensor node y_i to driver node u_j, $r_{ij} = \ell_{ij}$. Analogously, the relative degree $\rho_{i\kappa}$ can be defined as the geodesic

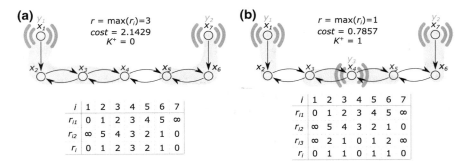

Fig. 3.1 Illustration of the concept of relative degrees and how one additional sensor decreases the relative degree of the system. **a** Relative degrees in the case of two sensors that are necessary to grant observability. **b** Relative degrees in the case of one additional sensor

path $\ell_{i\kappa}$ from sensor y_i to disturbance d_κ ($\rho_{i\kappa} = \ell_{i\kappa}$). The visual representation of the relative degree can be seen in Fig. 3.1.

When the aim of the observer is to estimate the effect of the $d(t) \in \Re^N$ disturbance vector that independently influences all state variables, then $\mathbf{E}$ is an identity matrix $\mathbf{I}_N$ and the relative degree of the system is $r = max_{i=1}^K r_i$. In the following, we assume that each state variable is influenced, thus relative degree is calculated between outputs and any of the state variables.

As the generation of a balanced placement of sensors is of interest, the set of sensor nodes S is determined by minimising the following cost functions:

$$\min_{S} cost(G, S, \beta) = \beta \max_{i=1}^K r_i + (1 - \beta)\frac{\sum_{i=1}^K r_i}{N} \tag{3.1}$$

where G stands for the network representation of the system, parameter $\beta = [0, 1]$ weights between the maximum and the balance-related average of the relative degree of the system. Besides the minimisation of the relative degree, the observability of the system should also be ensured. To generate the suitable output matrix $\mathbf{C}$ that fulfills these constraints, first the unmatched set of nodes S_f is generated by the path-finding method, followed by the set of candidate sensor nodes S_c. The cardinality of the candidate sensor nodes K^+ ($K^+ = |S_c|$) is the number of additional sensors. Then, matrix $\mathbf{C}$ is designed based on the resulted set of sensor nodes $S = S_f \cup S_c$ such that for each $x_i \in S$, $K = |S|$, c_j is a one-hot row vector whose ith element is non-zero, $j = 1, \ldots, K$, so $\mathbf{C} = [c_1^T, \ldots, c_K^T]^T$.

As the number of sensors which can be allocated without any restrictions is K^+, $\frac{K^+!}{|V \setminus S_f|!(|V \setminus S_f| - K^+!)}$ different combinations of possible placements of sensors exist. To solve this NP-hard problem, a centrality measures-based, a set covering-based and two simulated annealing-based heuristic optimisation algorithms are proposed in the following sections.

3.3 Fast and Robust Output Configuration Design

To optimise the relative degree and smoothness of the system, the objective function (Eq. 3.1), which can be considered as a linear combination of the maximum and average distance of the sensors from the observed state variables, was defined. In the following, three simple heuristic-based methods, namely the centrality measures-based and the set covering-based grassroot and retrofit methods utilise the maximum allowed relative degree r_{max}, while the second two simulated annealing and fuzzy clustering-based methods utilise the *cost* function during the optimisation.

3.3.1 Centrality Measures and Set Covering-Based Methods

Three methods were proposed to extend the minimal input and output configurations that use simple heuristics [20]. The first approach utilises two network-specific measures, namely the closeness and the node betweenness centrality measures. Firstly, the allowed maximal relative degree r_{max} is defined, and the set of necessary sensors (or drivers) is defined, O (or C). Then closeness centrality of node i is calculated using Eq. (3.2), where $N_R(i)$ is the number of reachable state variables from node i (node i excluded). If state variable j is unreachable from node i, then the distance between them is zero, $\ell_{i,j} = 0$. If node i cannot reach any of the state variable ($N_R(i) = 0$), then its distance form other state variables is considered as infinity, $\sum_{j \neq i} \ell_{i,j} = \infty$.

$$Cc(i) = \frac{N_R(i)^2}{(N-1) * \sum_{j \neq i} \ell_{i,j}} \tag{3.2}$$

Betweenness centrality calculates the number of geodesic paths that intercept node i, $\sigma_{st}(i)$, and divides this by the number of all the geodesic paths, σ_{st}, for each start s and target t node, such that $s \neq i \neq t$. The betweenness centrality is also normalised, i.e. if it is equal to one, then node i is part of each geodesic path, while when it is equal to zero, i is not part of any path. The betweenness centrality of node i is calculated as shown in Eq. (3.3).

$$Bc(i) = \sum_{s \neq i \neq t} \frac{\sigma_{st}(i)}{\sigma_{st}} \tag{3.3}$$

Then, for each sensor, e.g. for sensor i, the set of state variables Z_i is generated such that the closest sensor for the state variables in the set is sensor i. For each Z_i, the subgraph G_{Z_i} is generated that contains the nodes Z_i and all the edges exist between them in G. If the relative degree of G_{Z_i} exceeds r_{max}, a node with the highest combined centrality measure is selected as an additional sensor (or actuator), i.e., $O = O \cup \{i : \max(Cc(i)Bc(i)), i \notin O\}$ in the case of the output configuration and $C = C \cup \{i : \max (Cc(i)Bc(i)), i \notin C\}$ in the case of the input configuration. The steps are repeated iteratively until all G_{Z_i} exhibit a degree not in excessive of r_{max}. The pseudocode of the algorithm can be seen in Algorithm 1.

Algorithm 1 Pseudocode of the centrality measures-based method.

1: **procedure** CENMEASURES($\mathbf{A}$, O, r_{max})

2:      **for** $i = 1$ to $|O|$ **do**
3:          Let Z_i be the set of state variables closest for sensor i
4:          **if** $r_{max} < r_i$ **then**
5:              $O = O \cup \{i : \max (Cc(i)Bc(i)), i \notin O\}$
6:              $O = O \cup cenMeasures(Z_i, O, r_{max})$
7:          **end if**
8:      **end for**
9:      **return** O
10: **end procedure**

The second and third approaches use the set-covering method. They can be used for grassroot or retrofit design. Firstly, the set of nodes W_i reachable from node i over a maximum of r_{max} steps was determined. In the formulation of the algorithm, $\mathcal{U}$ denotes the set of all the state variables, O (or C) the set of the sensors (or actuators) necessary to ensure structural observability (or controllability). In the case of output (or input) configuration, J represents the set of necessary sensor (or driver) nodes, such that P is the set of state variables that is covered by J, namely $P = \cup_{j \in J} W_j$. The goal is to minimise J such that $P = \mathcal{U}$ and $O \subset J$ (or $C \subset J$); moreover, $r_u \leq r_{max}$, and $\forall u \in \mathcal{U}$. In case of grassroot design, when initial output (or input) configuration is not given, then the method yields a global optimum for the problem with respect to relative degree, but in this case structural observability (or controllability) is not granted. In the case of retrofit design, the state variables reachable from sensor (or driver) nodes are removed from W_i. A greedy algorithm was applied to solve the set-covering problem [15]. The pseudocode of the algorithm can be seen in Algorithm 2.

3.3.2 Simulated Annealing and Fuzzy Clustering-Based Methods

Based on this interpretation, the minimisation problem can also be seen as a k-medoid clustering problem, where the centroid elements are the sensor nodes and the members of the clusters are the observed state variables.

To determine the locations of the additional sensors two approaches are proposed inspired by Clustering Large Applications based on Simulated Annealing algorithm (CLASA) [10] and the Geodesic Distance-based Fuzzy c-Medoid Clustering method (GDFCM) [18]. In the first algorithm, the CLASA algorithm is modified as follows.

Algorithm 2 Pseudocode of the set covering-based method.

1: **procedure** SETCOVERING($\mathbf{A}, r_{max}, O$)
2:     **for** $i = 1$ to N **do**
3:         Let W_i be the set of state variables reachable from node i in r_{max} steps
4:     **end for**
5:     **if** $O \neq \emptyset$ **then**
6:         $P = \cup_{o \in O} W_o$
7:         **for** $i = 1$ to N **do**
8:             $W_i = W_i \setminus P$
9:         **end for**
10:     **end if**
11:     $J = O \cup greedyScp(W_i)$ ▷ see [15]
12:     **return** J
13: **end procedure**

The original objective function of CLASA that calculates the distances between the objects and the medoids (in our case the distances between the state variables and the sensors) is replaced by what was proposed in Eq. 3.1. With the new objective function, not just the minimum relative degree can be granted, but different sensor configurations with the same relative degree can be distinguished to balance the load of the sensors. The search mechanism should be fine-tuned as the medoids of the fixed sensors, S_f, have fixed positions to ensure the observability of the system. The second algorithm enhances the random search of the resulted mCLASA algorithm by the introduction of distance-dependent selection probability calculated based on the Geodesic Distance-based Fuzzy c-Medoid Clustering method (GDFCM) [18].

In both algorithms, firstly the S_f fixed sensor nodes are determined to grant the observability property based on the unmatched set of nodes. Then the candidate sensor nodes S_c are generated randomly, and the cost is calculated. Following this in each iteration, one of the candidate sensor nodes, s^-, is randomly changed, and the new cost is calculated. From this, the difference between the new cost and the previously applied cost is calculated, Δ. The new placement of the sensors is accepted if $\Delta < 0$.

If $\Delta \geq 0$, the poorer placement of the sensor is only accepted with the probability defined as $\exp(\frac{-\Delta}{T^i})$, where T^i denotes the temperature of the cooling process of the SA in iteration i, the dynamics of which can be seen in Eq. 3.4, where α is the reduction rate of the temperature that can initially be a defined value or can be determined for a given number of iterations, $maxiter$, as well as maximum and minimum temperatures, T_{max} and T_{min}. This randomisation of the search allows that the algorithm to explore the search space and converge with the increase of the number of iterations.

$$T^{i+1} = \alpha T^i, \qquad\qquad \alpha = \left(\frac{T_{min}}{T_{max}}\right)^{1/maxiter} \qquad (3.4)$$

The pseudocode of the proposed modified clustering large applications based on simulated annealing algorithm (mCLASA) can be seen in Algorithm 3. The algorithm possesses the following inputs and parameters: G denotes the network representation of the system from Eq. 1.1, $\beta \in (0, 1)$ stands for the parameter of the cost functions from Eq. 3.1, K^+ represents the number of additional sensors, $maxiter$ is the number of iterations of the simulated annealing, T_{max} denotes the maximum, i.e. the initial temperature and T_{min} stands for the minimum temperature. The following values were used for the parameters: $\beta = 0.5$, $maxiter = 500$, $T_{max} = 10$ and $T_{min} = 10^{-2}$.

Algorithm 3 Pseudocode of the modified CLASA algorithm (mCLASA).

1: **procedure** MCLASA(G, β, K^+, $maxiter$, T_{max}, T_{min})

2:      $\alpha = \left(\frac{T_{min}}{T_{max}}\right)^{1/maxiter}$, $T^1 = T_{max}$

3:      $S_f = getSensors(G)$ ▷ using the path-finding method: [19]

4:      Let S_c be the set of the K^+ randomly chosen elements from $V \backslash S_f$

5:      $cost = cost(G, S_f \cup S_c, \beta)$ ▷ Equation 3.1

6:      **for** $i = 1$ to $maxiter$ **do**

7:          Let s^- be a randomly selected element from S_c

8:          $S'_c = S_c \backslash \{s^-\}$

9:          Let s^+ be a randomly chosen node from $V \backslash (S_f \cup S'_c)$

10:         $S'_c = S'_c \cup \{s^+\}$

11:         $newcost = cost(G, S_f \cup S'_c, \beta)$

12:         $\Delta = newcost - cost$

13:         **if** $\Delta < 0$ **then**

14:             $cost = newcost$

15:             $S_c = S'_c$

16:         **else**

17:             **if** $random() < \exp(\frac{-\Delta}{T^i})$ **then**

18:                 $cost = newcost$

19:                 $S_c = S'_c$

20:             **end if**

21:         **end if**

22:         $T^{i+1} = \alpha T^i$

23:     **end for**

24:     **return** $cost$

25: **end procedure**

The second method utilises the GDFCM method [18] to extend the previously introduced mCLASA method. This extension increases the probability that in each iteration potentially better neighbouring sensors are selected to replace the sensors placed in the previous iteration steps, s^-. The algorithm considers the sensors as cluster centres (medoids), determines the cluster assignments of the state variables and swaps the medoid (the sensor) with a randomly selected state variable. The random selection utilises distance-based fuzzy membership values. The cluster memberships $\mu_{i,j}$ of the state variables $R(s_j)$ are determined by the fuzzy membership function (see Equation (3.5) in the description of the algorithm), where the distances are calculated in the undirected version of the graph (G^u). A roulette wheel selection [14] selects the new medoid according to the membership values. The search space of the algorithm is also controlled by the fuzzy exponent m which is decreased in each iteration by α_m, in the same manner as α in Eq. 3.4.

The pseudocode of the suggested Geodesic Distance-based Fuzzy c-medoid Clustering method with Simulated Annealing algorithm (GDFCMSA) can be seen in Algorithm 4. The algorithm possesses the following inputs and parameters: G denotes the network representation of the system, $\beta \in (0, 1)$ stands for the balance parameter of the cost functions, K^+ represents the number of additional sensors, $maxiter$ is the number of iterations of the simulated annealing, m_{max} denotes the maximum fuzzy component value and m_{min} stands for the minimum fuzzy component value. The following values were used for the parameters: $\beta = 0.5$, $maxiter = 500$, $T_{max} = 10$, $T_{min} = 10^{-2}$, $m_{max} = 10$ and $m_{min} = 2$.

Both mCLASA and GDFCMSA methods can be parameterised such that the maximum acceptable relative degree, r_{max} is given instead of the number of additional nodes, K^+. In this case, a new cycle contains the whole algorithm presented in Algorithm 3 or 4, where the cycle repeats until the resulted maximum relative degree is smaller or equal than r_{max} and in each iteration K^+ is increased by 1.

3.4 Sensor Placement Case Studies

In this section, four case studies will be introduced to demonstrate the applicability of the previously presented methods. After a brief introduction of the case studies, we evaluate the proposed methods from three aspects. Firstly, we analyse the performance of the methods based on how they can ensure an initially defined demanded relative degrees; secondly, the effect of the additional sensor nodes on the cost function and to the relative degree of the system is studied; thirdly the speed of converge of SA-based methods is analysed.

3.4.1 Description of the Case Studies

The first case study is related the control of heat exchanger networks (HENs). HENs are widely studied dynamical systems because the complexity of interlinked heat exchangers requires advanced process monitoring and control algorithms. The network topology of the studied HEN is shown in Fig. 3.2. The network consists of six hot streams, two cold streams, ten heat exchanger cells and two utility coolers [36]. The state-transition matrix $\mathbf{A}$ of the problem can be easily determined based on the structure of the process [34], which results in results in 22 state variables.

Following the detailed analysis of the problem with regard to the placement of sensors in the HEN, three other dynamical systems will be analysed to illustrate the applicability of the methods on larger examples. These examples are not typical control-relevant problems as they are used to study model-reduction algorithms [7],

---

Algorithm 4 Pseudocode of the Geodesic Distance-based Fuzzy c-Medoid Clustering method with Simulated Annealing algorithm (GDFCMSA).

1: **procedure** GDFCMSA($G, \beta, K^+, maxiter, T_{max}, T_{min}, m_{max}, m_{min}$)

2:     $\alpha = \left(\frac{T_{min}}{T_{max}}\right)^{1/maxiter}, T^1 = T_{max}, \alpha_m = \left(\frac{m_{min}}{m_{max}}\right)^{1/maxiter}, m^1 = m_{max}$

3:     $S_f = getSensors(G)$ ▷ using the path-finding method: [19]

4:     Let S_c be the set of the K^+ randomly chosen state elements from $V \setminus S_f$

5:     $cost = cost(G, S_f \cup S_c, \beta)$ ▷ Equation 3.1

6:     **for** $i = 1$ to $maxiter$ **do**

7:         Calculate the fuzzy membership values for G^u, $1 \le i \le K$ and $1 \le j \le N$:

$$\mu_{i,j} = \frac{1}{\sum_{k=1}^{K}\left(\frac{r_{i,j}}{r_{k,j}}\right)^{\frac{2}{m^i-1}}} \qquad (3.5)$$

8:         Let s^- be a randomly selected sensor node from S_c

9:         $S_c' = S_c \setminus \{s^-\}$

10:        $R(s^-) = \{x_j | x_j \in V, s^- = \arg \max_{k=1}^{K} \mu_{k,j}\}$

11:        **for all** $s_j \in R(s^-)$ **do**

12:            $P(s_j) = \mu_{s^-, s_j}$

13:        **end for**

14:        $s^+ = \{x_j | x_j = roulette\_wheel\_selection(P)\}$

15:        $S_c' = S_c' \cup \{s^+\}$

16:        $newcost = cost(G, S_f \cup S_c', \beta)$

17:        $\Delta = newcost - cost$

18:        **if** $\Delta < 0$ **then**

19:            $cost = newcost$

20:            $S_c = S_c'$

21:        **else**

22:            **if** $random() < \exp(\frac{-\Delta}{T^i})$ **then**

23:                $cost = newcost$

24:                $S_c = S_c'$

25:            **end if**

26:        **end if**

27:        $T^{i+1} = \alpha T^i, m^{i+1} = \alpha_m m^i$

28:    **end for**

29:    **return** $cost$

30: **end procedure**

---

but these problems illustrate how the algorithms are able to detect structurally efficient placements. The first two additional examples are based on Modified Nodal Analysis (MNA) [26] and contain 578 and 980 state variables, respectively. The last case study is based on the state- space model of a partial element equivalent circuit (PEEC) of a patch antenna structure with 172 inductances, 6990 mutual inductances and 2100 capacitances that define 480 state variables [16]. Further information about the case studies can be found in Sect. B.1 of the Appendix.

The performance analysis of the case studies contains three scenarios. Firstly, for an initially determined set of r_{max} the resulted costs, balances of relative degrees and

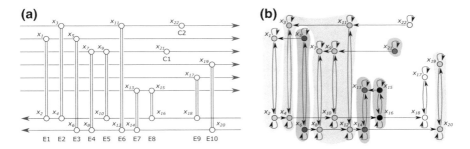

Fig. 3.2 a The structure of the studied Heat Exchanger Network represents how cold and hot streams influence each other. **b** The network representation of the state variables are related to the temperatures of the streams depicted in Fig. 3.2a. Colours indicate strongly connected components

Table 3.1 The main parameters of the networks studied in the examples. $|V|$ denotes the number of state variables, $|E|$ stands for the number of the connections between the state variables, $|S_f|$ represents the number of fixed sensor nodes, K^+ denotes the number of additional nodes to be analysed and r_{max} stands for the maximum acceptable relative degrees used as parameters

Network	Description	$|V|$	$|E|$	$|S_f|$	K^+	r_{max}
HEN	A simple heat exchanger network	22	56	3	$0, \ldots, 15$	$\{5,4,3,2,1\}$
MNA_1	Modified Nodal Analysis of a multiport, voltage sources	578	1,694	3	$0, \ldots, 35$	$\{60,50,40,30,20\}$
MNA_4	Modified Nodal Analysis of a multiport, voltage sources	980	2,872	2	$0, \ldots, 35$	$\{89,75,60,45,30\}$
PEEC	Partial element equivalent circuit model	480	1,346	1	$0, \ldots, 35$	$\{54,40,30,20,10\}$

number of assigned sensors are analysed for all the four proposed methods. Secondly, as mCLASA and GDFCMSA method can be parameterised with K^+ instead of r_{max}, the effect of the additional sensors on cost and relative degree is analysed in the case of these two methods. Thirdly, the convergence of the two simulated annealing-based methods is analysed. The assigned parameters of the case studies are summarised in Table 3.1. The network representations of the case studies and the resulted sensor placements are presented in Appendix B.2.

In the cases of application of mCLASA and GDFCMSA methods, as they are random search-based techniques, each result was evaluated based on 100 independent runs of the algorithms. The parameters of the algorithms were the same in each scenario: $\beta = 0.5$, $T_{max} = 10$, $T_{min} = 10^{-2}$, $m_{max} = 10$, $m_{min} = 2$. The reduction rates α and α_m were determined by Eq. 3.4, as can be seen in Algorithms 3 and 4.

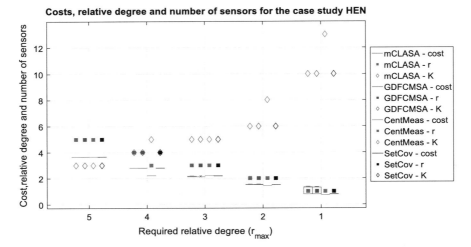

Fig. 3.3 The values of the cost function, relative degree and number of outputs with regard to the maximum allowed relative degree r_{max} in the case studies HEN. The box plots show the interquartile ranges and the medians of the cost function, while the scatter plots with square visualise the resulted relative degree. Scatter plots with diamond shape denote the number of sensors placed in the system. The results of mCLASA, GDFCMSA, centrality measures-based and set- covering-based methods are presented in colour red, blue, magenta and black, respectively. Plus signs denote the outliers (if they exist) of the corresponding boxes

3.4.2   Performance of the Optimisation Algorithms

In the first scenario, firstly, the S_f set of fixed sensors that are required to ensure the observability of the system is determined. Then, in each iteration an additional sensor added to the system to reduce relative degree, r. To evaluate the robustness of the random search-based algorithms the results of 100 independent runs were visualised in box plots to show the distribution of the optimised cost functions. The first column in each result (Figs. 3.3, 3.4, 3.5 and 3.6) is the result provided by maximum matching, i.e. r_{max} is equal with the relative degree granted by minimal configurations.

The methods mCLASA and GDFCMSA propose nearly the same results. Although the centrality measures-based method can provide a better result according to relative degree r and $cost$ function, the number of additional nodes, K^+ is higher. The advantage of simulated annealing can be seen as the result of mCLASA and GDFCMSA is better than in case of set covering-based method which usually assigns the same amount of additional sensor to the system.

To compare the performance of the methods, a larger dataset was also utilised, that is introduced in Sect. 4.3. The dataset contains more than 600 Heat Exchanger Networks. For each network, we set r_{max} as part of the quarter of the diameter of the

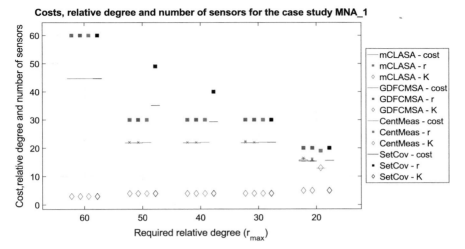

Fig. 3.4 The values of the cost function, relative degree and number of outputs with regard to the maximum allowed relative degree r_{max} in the case studies MNA_1. The box plots show the interquartile ranges and the medians of the cost function, while the scatter plots with square visualise the resulted relative degree. Scatter plots with diamond shape denote the number of sensors placed in the system. The results of mCLASA, GDFCMSA, centrality measures-based and set covering-based methods are presented in colour red, blue, magenta and black, respectively. Plus signs denote the outliers (if they exist) of the corresponding boxes

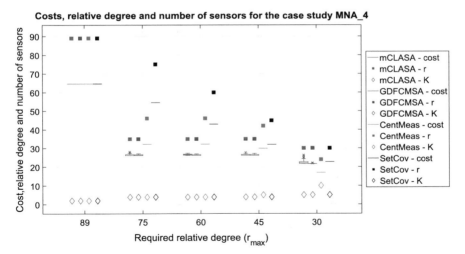

Fig. 3.5 The values of the cost function, relative degree and number of outputs with regard to the maximum allowed relative degree r_{max} in the case studies MNA_4. The box plots show the interquartile ranges and the medians of the cost function, while the scatter plots with square visualise the resulted relative degree. Scatter plots with diamond shape denote the number of sensors placed in the system. The results of mCLASA, GDFCMSA, centrality measures-based and set covering-based methods are presented in colour red, blue, magenta and black, respectively. Plus signs denote the outliers (if they exist) of the corresponding boxes

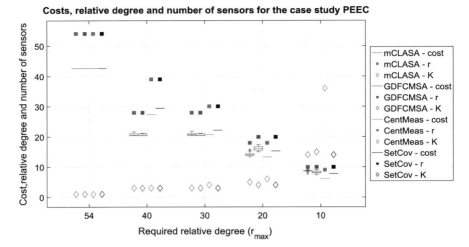

Fig. 3.6 The values of the cost function, relative degree and number of outputs with regard to the maximum allowed relative degree r_{max} in the case studies PEEC. The box plots show the interquartile ranges and the medians of the cost function, while the scatter plots with square visualise the resulted relative degree. Scatter plots with diamond shape denote the number of sensors placed in the system. The results of mCLASA, GDFCMSA, centrality measures-based and set covering-based methods are presented in colour red, blue, magenta and black, respectively. Plus signs denote the outliers (if they exist) of the corresponding boxes

Table 3.2 Result of the comparison of the methods. In row i, column j the number of HENs is presented, where method i ensures r_{max} with fewer sensors than method j

	CentMeas	SetCov	mCLASA	GDFCMSA
CentMeas	0	0	2	1
SetCov	523	0	69	45
mCLASA	528	43	0	17
GDFCMSA	531	61	66	0

network. Then, we analyse that how many times a method grants a better solution by the number of sensors assigned. The result can be seen in Table 3.2.

Besides the better solution, the equally good solutions were examined, i.e. where the two methods have the same performance level by the number of sensors assigned. The number of HENs, where the result of the comparison is tie is presented in Table 3.3.

Last, but not least in the first scenario, Pareto frontier was generated. Two cases were determined. Firstly, we evaluate the methods as how fine is the granted solution based on the provided cost function, and the balance of relative degree. In this case, a two-dimensional plane was stretched by the methods. Secondly, as third dimension, we added the number of determined sensors to the evaluation. This results in more method on the front, but the results are still valuable, that can be seen in Table 3.4. The

Table 3.3 Result of the comparison of the methods, where the methods proposed equally good configurations. In row i, column j the number of HENs is presented, where method i and j ensure r_{max} with the same number of sensors

	CentMeas	SetCov	mCLASA	GDFCMSA
CentMeas	0	116	109	107
SetCov	116	0	527	533
mCLASA	109	527	0	556
GDFCMSA	107	533	556	0

Table 3.4 Pareto frontiers that were generated to 639 HENs. For each method, the number when the method was a member of the frontier is represented. In the 2-dimensional case, the frontier was generated based on the *cost* function and the balance of relative degrees, while in the 3-dimensional case, the number of sensor nodes was taken into account as the third dimension besides the previous two

	CentMeas	SetCov	mCLASA	GDFCMSA
2-dimensional	622	605	614	612
3-dimensional	631	638	638	639

centrality measures-based method is highly the best in 2-dimensional Pareto frontier, but the consideration of the number of assigned sensors can destroy its advantage. However, based on Tables 3.2 and 3.3 the advantage of simulated annealing is still unequivocal.

In the second scenario, firstly, the S_f set of fixed sensors that are required to ensure the observability of the system is determined. The number of sensors added to improve the dynamical properties of the system is represented by the cardinality of $|K^+|$.

To evaluate the robustness of the random search-based algorithms the results of 100 independent runs were visualised in box plots to show the distribution of the optimised cost functions. As can be seen in Figs. 3.7 and 3.8, the algorithms are consistent in most of the cases; however, the presence of outliers show that the algorithms can fail in local optimums.

It is visible that in all cases both the cost functions and the relative degrees of the observed state variables decrease significantly as the number of sensors increases, so a relatively small quantity of additional sensors can cause a noticeable improvement in terms of the relative degree compared to the observer designed only to ensure observability. It can also be pointed out that additional sensors ($K^+ = 25$ or 35) do not significantly reduce the relative degree of the system.

The heights of the box plots show the interquartile ranges of the optimised cost functions. Higher box plots reflect higher uncertainty in the solutions which could indicate the increased difficulty of the sensor placement problem. For example, in Fig. 3.7, for $K^+ = 7$ the box plot shows a much wider distribution than for other K^+ that can be explained there is a transient in the structure of the solutions.

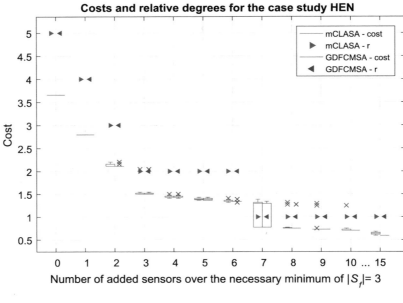

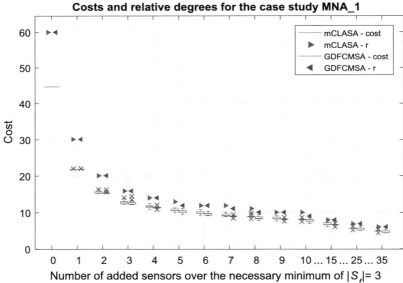

Fig. 3.7 The values of the cost functions and relative degrees with regard to the number of added sensors in the case studies HEN and MNA_1. The box plots show the interquartile ranges and the medians of the cost functions, while the scatter plots visualise the minimum value of the relative degree for 100 independent runs. For each value of the x axis, the results provided by the mCLASA are shown on the left-hand side, and the results provided by the GDFCMSA algorithm are shown on the right-hand side. The triangles represent the relative degrees, while the plus signs denote the outliers (if they exist) of the corresponding boxes

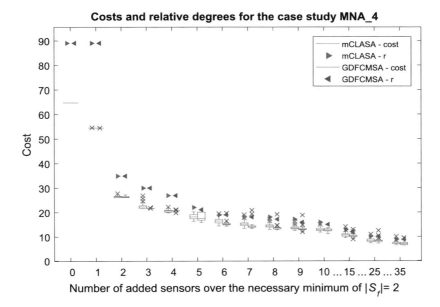

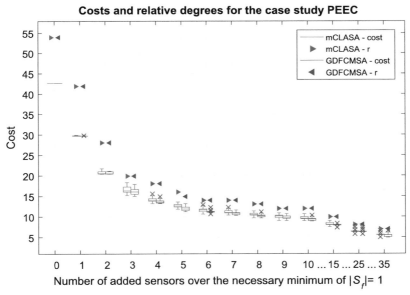

Fig. 3.8 The values of the cost functions and relative degrees with regard to the number of added sensors in the case studies MNA_4 and PEEC. The box plots show the interquartile ranges and the medians of the cost functions, while the scatter plots visualise the minimum value of the relative degree for 100 independent runs. For each value of the x axis, the results provided by the mCLASA are presented on the left-hand side, while the results provided by the GDFCMSA algorithm are shown on the right-hand side. The triangles denote the relative degrees, while the plus signs represent the outliers of the corresponding boxes

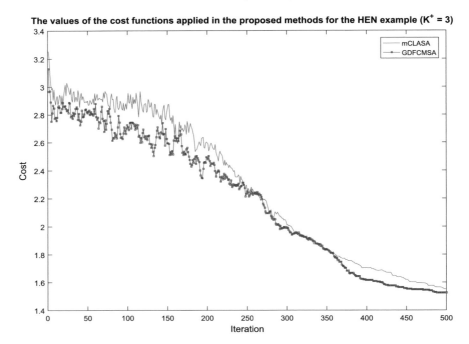

Fig. 3.9 Analysis of the convergence with regard to the proposed methods in the case of the HEN. The colour red denotes the method mCLASA while the colour blue denotes the method GDFCMSA. It is clearly visible that the value of *maxiter* should be high because the methods also converge after 500 iterations. GDFCMSA provides a slightly better solution

3.4.3 Convergence Analysis

In the third scenario, the converge of the proposed methods was tested. In Fig. 3.9, the convergences of the methods can be seen for the HEN case study with the parameter setting $K^+ = 3$. The results and the network representation of the other case studies are presented in Appendix B.3. In the figure, it can be seen how fast the methods converge in terms of the number of iterations. With the addition of three sensors, the cost functions was reduced by more than 50% over the 500 iterations which illustrate well the benefit of the proposed methods.

In this figure, it can also be seen, that it is worth selecting a high value of the parameter *maxiter* because even during the 500th iteration the cost functions show still show a small decrease.

From this figure, it is evident, that the fuzzy extension of the algorithm improves the speed of convergence which is also reflected from the slightly better performances shown in Figs. 3.7 and 3.8.

The durations of the calculations were also measured with *maxiter* = 500 to provide an illustrative comparison between the algorithm and the complexity of the

Table 3.5 Durations of calculations in seconds in terms of the optimal placement of K^+ additional sensors. The presented results are the means of 100 independent runs

K^+	HEN		MNA_1		MNA_4		PEEC	
	mCLASA	GDFCMSA	mCLASA (s)	GDFCMSA (s)	mCLASA (s)	GDFCMSA (s)	mCLASA (s)	GDFCMSA (s)
1	0.0158 s	0.0164 s	0.1311	0.1396	0.1795	0.1892	0.0704	0.0715
2	0.0173 s	0.0183 s	0.1752	0.1689	0.2240	0.2306	0.0945	0.0955
3	0.0205 s	0.0227 s	0.2230	0.2120	0.2676	0.2922	0.1176	0.1179
5	0.0233 s	0.0280 s	0.2638	0.2999	0.3724	0.3749	0.1602	0.1614
10	0.0310 s	0.0315 s	0.3965	0.4106	0.6353	0.7106	0.2829	0.2850
15	0.0427 s	0.0428 s	0.5984	0.6721	0.9049	0.9176	0.4631	0.4237
25	n.a.	n.a.	0.8052	0.8170	1.5417	1.4754	0.6385	0.6862
35	n.a.	n.a.	1.3258	1.3404	2.1008	2.0255	0.9229	1.0077

problem. The results are presented in Table 3.5. The durations of calculations are valid for our notebook environment using Windows 10, an Intel Core i7-6600U processor, 16 GB of RAM and MATLAB R2016a. These results can be reached if the MatlabBGL [13] implementation of searching all the shortest paths is used in NOCAD. With this modification, the Octave compatibility is lost unfortunately.

The short running times shown in Table 3.5 arise the question whether the exhaustive enumeration of the solutions could handle the problem, is the heuristic search is necessary? The problem is combinatorially complex, for each K^+ $\binom{N-|S_f|}{K^+}$ possible solutions should be examined. In the HEN case study for $K^+ = 9$, the SA examines only 500 solutions instead of the 92,378 possible configurations. For larger problem the difference is more significant, e.g. the example of MNA_4 with $K^+ = 35$ would require $\binom{978}{35} \approx 2.4 \times 10^{64}$ function evaluations although this problem neither can be considered as a huge system.

3.5 Discussion

To ensure the structural observability in addition to fast and robust observer response, three simple heuristic- and two clustering- and simulated annealing-based methods were proposed.

To decrease the relative degree of the designed system, centrality measures-based method is proposed that utilises the closeness and betweenness centrality measures to determine additional sensors for the output configurations. Beside centrality measures, set covering is also utilised to determine the necessary additional sensors.

Additional sensors are also placed into the system based on the CLASA algorithm by the mCLASA algorithm. The placements of additional sensors are further improved by the Geodesic Distance-based Fuzzy c-Medoid Clustering with Simulated Annealing algorithm (GDFCMSA) as their positioning is based on a geodesic distance-based fuzzy membership functions. Simulated annealing is applied by both algorithms to minimise the cost functions, which in turn minimises the maximum and average of the relative degrees of the system to generate a balanced placement of sensors.

A slightly better solution is provided by the GDFCMSA algorithm than the mCLASA method at the expense of some additional computational resources for the evaluation of the membership degrees and both of them provide a better solution than centrality measures and set covering-based methods. Solutions to massive problems are generated by both methods over a short period of time; which illustrates the applicability of these methods in an industrial setting.

Although this chapter deals with the placement of sensors, the proposed methodology can be applied to control configuration design as well, since this is the dual problem of the design of the output configuration. From this viewpoint, these methods can be utilised in a more diverse series of problems, e.g. the minimisation of the time necessary to spread information in a social network by influencers, or the optimisation of a transportation network to decrease journey times. As the benchmark examples are questionable as introduced in Chap. 2, the HEN presented here is a well-defined, real dynamical system that can be a valuable basis of network science-based analysis. Thus, in the next chapter, we analyse the HENs that can be found in the literature to make statements about how structural analysis can be applied in the prediction of dynamical behaviours.

References

1. Aguirre-Salas, L., Begovich, O., Ramirez-trevino, A.: Sensor assignment for observability in interpreted petri nets. IFAC Proc. Vol. **37**(18), 441–446 (2004)
2. Bagajewicz, M.J.: Design and retrofit of sensor networks in process plants. AIChE J. **43**(9), 2300–2306 (1997)
3. Berger-Wolf, T.Y., Hart, W.E., Saia, J.: Discrete sensor placement problems in distribution networks. Math. Comput. Model. **42**(13), 1385–1396 (2005)
4. Blanloeuil, P., Nurhazli, N.A.E., Veidt, M.: Particle swarm optimization for optimal sensor placement in ultrasonic shm systems. In: Nondestructive Characterization and Monitoring of Advanced Materials, Aerospace, and Civil Infrastructure 2016, vol. 9804, p. 98040E. International Society for Optics and Photonics (2016)
5. Boukhobza, T., Hamelin, F.: State and input observability recovering by additional sensor implementation: a graph-theoretic approach. Automatica **45**(7), 1737–1742 (2009)
6. Carballido, J.A., Ponzoni, I., Brignole, N.B.: Cgd-ga: a graph-based genetic algorithm for sensor network design. Inf. Sci. **177**(22), 5091–5102 (2007)
7. Chahlaoui, Y., Van Dooren, P.: A collection of benchmark examples for model reduction of linear time invariant dynamical systems (2002)
8. Chiu, P.L., Lin, F.Y.S.: A simulated annealing algorithm to support the sensor placement for target location. In: 2004 Canadian Conference on Electrical and Computer Engineering, vol. 2, pp. 867–870. IEEE (2004)
9. Chmielewski, D.J., Palmer, T., Manousiouthakis, V.: On the theory of optimal sensor placement. AIChE J. **48**(5), 1001–1012 (2002)

10. Shu-Chuan, C., Roddick, J.F., Pan, J-S.: A comparative study and extension to K-medoids algorithms, Contemporary Development Company (2001)
11. Daoutidis, P., Kravaris, C.: Structural evaluation of control configurations for multivariable nonlinear processes. Chem. Eng. Sci. **47**(5), 1091–1107 (1992)
12. Düştegör, D., Frisk, E., Cocquempot, V., Krysander, M., Staroswiecki, M.: Structural analysis of fault isolability in the damadics benchmark. Control Eng. Pract. **14**(6), 597–608 (2006)
13. Gleich, D.: Matlabbgl. a matlab graph library. Institute for Computational and Mathematical Engineering, Stanford University (2008)
14. Goldberg, D.E.: Genetic Algorithms in Search, Optimization, and Machine Learning. Addison-Wesley (1989)
15. Gori, F., Folino, G., Jetten, M.S.M., Elena Marchiori. Mtr: taxonomic annotation of short metagenomic reads using clustering at multiple taxonomic ranks. Bioinformatics **27**(2), 196–203 (2010)
16. Heeb, H., Ruehli, A.E., Eric Bracken, J., Rohrer, R.A.: Three dimensional circuit oriented electromagnetic modeling for vlsi interconnects. In: IEEE 1992 International Conference on Computer Design: VLSI in Computers and Processors, 1992. ICCD'92. Proceedings, pp. 218–221. IEEE (1992)
17. Isidori, A.: Nonlinear Control Systems. Springer Science & Business Media, Berlin (2013)
18. Király, A., Vathy-Fogarassy, Á., Abonyi, J.: Geodesic distance based fuzzy c-medoid clustering-searching for central points in graphs and high dimensional data. Fuzzy Sets Syst. **286**, 157–172 (2016)
19. Leitold, D., Vathy-Fogarassy, Á., Abonyi, J.: Controllability and observability in complex networks-the effect of connection types. Sci. Rep. **7**, 151 (2017)
20. Leitold, D., Vathy-Fogarassy, A., Abonyi, J.: Design-oriented structural controllability and observability analysis of heat exchanger networks. Chem. Eng. Trans. **70**, 595–600 (2018)
21. Leitold, D., Vathy-Fogarassy, Á., Abonyi, J.: Network distance-based simulated annealing and fuzzy clustering for sensor placement ensuring observability and minimal relative degree. Sensors **18**(9), 3096 (2018)
22. Liu, X., Mo, Y., Pequito, S., Sinopoli, B., Kar, S., Aguiar, A.P.: Minimum robust sensor placement for large scale linear time-invariant systems: a structured systems approach. IFAC Proc. Vol. **46**(27), 417–424 (2013)
23. Liu, Y.-Y., Slotine, J.-J., Barabási, A.-L.: Observability of complex systems. Proc. Natl. Acad. Sci. **110**(7), 2460–2465 (2013)
24. Mamano, N., Hayes, W.: Sana: simulated annealing network alignment applied to biological networks (2016). arXiv:1607.02642
25. Martin, O.C., Otto, S.W.: Combining simulated annealing with local search heuristics. Ann. Oper. Res. **63**(1), 57–75 (1993)
26. Odabasioglu, A., Celik, M., Pileggi, L.T.: Prima: passive reduced-order interconnect macro-modeling algorithm. In: Proceedings of the 1997 IEEE/ACM International Conference on Computer-Aided Design, pp. 58–65. IEEE Computer Society (1997)
27. Papadimitriou, C., Lombaert, G.: The effect of prediction error correlation on optimal sensor placement in structural dynamics. Mech. Syst. Signal Process. **28**, 105–127 (2012)
28. Papadopoulou, M., Raphael, B., Smith, I.F.C., Sekhar, C.: Hierarchical sensor placement using joint entropy and the effect of modeling error. Entropy **16**(9), 5078–5101 (2014)
29. Qin, B.Y., Lin, X.K.: Optimal sensor placement based on particle swarm optimization. In: Advanced Materials Research, vol. 271, pp. 1108–1113. Trans Tech Publications (2011)
30. Rosich, A., Sarrate, R., Puig, V., Escobet, T.: Efficient optimal sensor placement for model-based fdi using an incremental algorithm. In: 46th IEEE Conference on Decision and Control, pp. 2590–2595. IEEE (2007)
31. Samora, I., Franca, M.J., Schleiss, A.J., Ramos, H.M.: Simulated annealing in optimization of energy production in a water supply network. Water Res. Manag. **30**(4), 1533–1547 (2016)
32. Padula, S.L.: and Kincaid Rex K. Optimization strategies for sensor and actuator placement. Technical report, National Aeronautics and Space Administration, NASA (1999)

33. Torres-Jimenez, J., Izquierdo-Marquez, I., Garcia-Robledo, A., Gonzalez-Gomez, A., Bernal, J., Kacker, R.N.: A dual representation simulated annealing algorithm for the bandwidth minimization problem on graphs. Inf. Sci. **303**, 33–49 (2015)

34. Varga, E.I., Hangos, K.M., Szigeti, F.: Controllability and observability of heat exchanger networks in the time-varying parameter case. Control Eng. Pract. **3**(10), 1409–1419 (1995)

35. Watson, J-P., Hart, W.E., Berry, J.W.: Scalable high-performance heuristics for sensor placement in water distribution networks. In: *Impacts of Global Climate Change*, pp. 1–12. American Society of Civil Engineers (2005)

36. Westphalen, D.L., Young, B.R., Svrcek, W.Y.: A controllability index for heat exchanger networks. Ind. Eng. Chem. Res. **42**(20), 4659–4667 (2003)

37. Worden, K., Burrows, A.P.: Optimal sensor placement for fault detection. Eng. Struct. **23**(8), 885–901 (2001)

38. Yuen, K.-V., Kuok, S.-C.: Efficient bayesian sensor placement algorithm for structural identification: a general approach for multi-type sensory systems. Earthq. Eng. Struct. Dyn. **44**(5), 757–774 (2015)

39. Zhang, X.-X., Li, H.-X., Qi, C.-K.: Spatially constrained fuzzy-clustering-based sensor placement for spatiotemporal fuzzy-control system. IEEE Trans. Fuzzy Syst. **18**(5), 946–957 (2010)

40. Xun, Z., Juelong, L., Jianchun, X., Ping, W., Qiliang, Y., Ronghao, W., Can, H.: Optimal sensor placement for latticed shell structure based on an improved particle swarm optimization algorithm. Math. Probl. Eng. 2014 (2014)

Chapter 4
Application to the Analysis of Heat Exchanger Networks

Abstract This work proposes a network science-based analysis tool for the qualification of controllability and observability of HENs. With the proposed methodology, the main characteristics of HEN design methods are determined, the effect of structural properties of HENs on their dynamical behaviour is revealed, and the potentials of the network-based HEN representations are discussed. Our findings are based on the systematic analysis of almost 50 benchmark problems related to 20 different design methodologies.

Keywords Heat exchanger network · Structural controllability · Structural observability · Operability · Network science · Sensor and actuator placement

4.1 Importance of Process Integration

More now than ever, industrial processes are integrated to increase efficiency [11]. Process Integration (PI) dates back to 1970 when PI was the response to the oil crisis, and, since then, this field has been a hot topic as it can be utilised to minimise energy and water usage [52], waste as well as emissions [26]. Since then, tools developed for Heat and Energy Integration have become essential elements of Process Design [27], including Energy Storage Systems [24]. Meanwhile, PI increases the efficiency of the technologies, and the increased complexity also complicates operations. Heat Integration also often results in more complex but less operable technologies [25]. Thus, a methodology to support the integrated system design must have a good qualitative and quantitative model that highlights the characteristics of the processes [60].

As we presented in Chap. 1, many works deal with the controllability [38] and observability [39] of complex systems, but the connection between the complexity of the system and the difficulty of operations has not been examined in details. However, the analysis of the structural motifs of the building elements of complex systems and their models can highlight potentially useful information that can support the design and operation. The structure-relevant analysis of heat exchanger networks (HENs) has already been proven to be beneficial, e.g. the resilience index

© The Author(s), under exclusive license to Springer Nature Switzerland AG 2020   49
D. Leitold et al., *Network-Based Analysis of Dynamical Systems*,
SpringerBriefs in Computer Science,
https://doi.org/10.1007/978-3-030-36472-4_4

(RI) quantifies the ability of a HEN to deal with disturbances [46], the controllability index measures the controllability of the HENs [56], and structural analysis can be applied to determine the locations of additional sensors that can reveal otherwise indistinguishable faults [17].

The connection between the structural properties of HENs designed by different methodologies and the operability of the system is yet to be examined in detail. The optimisation of HENs can be formalised in several ways. Algorithms can focus on the minimum number of matches [33], the maximum energy recovery (MER) [34], the minimum energy–capital cost [2], the minimum total annualised cost (TAC) [5], the minimum number of exchangers [42], or, in the case of retrofit design, the minimum number of additional exchangers and the additional area of the exchangers or piping costs [12]. The goals listed above can be achieved by different algorithms, such as the Pinch methodology [36], dual-temperature approach method [50], pseudo-pinch [57], supertargeting [34], state-space approach [5], branch- and bound-based algorithms [44], or the application of genetic algorithm and simulated annealing (GA and SA, respectively) [59].

Although there are no systematic studies related to how optimisation algorithms affect the structural properties of HENs with regard to operability, there are some well-known relations. The pinch methodology determines the minimum temperature difference, ΔT_{min}, which divides the HEN into two or more subnetworks: above and below the Pinch points [36]. The method enables an MER design to be created, i.e. it minimises the energy required. The price of the MER design is the increased number/area and installation costs of heat exchangers. In contrast, optimisation based on the minimum number of matches decreases the number of heat exchangers in addition to the installation cost [29], and makes the operation easier [42]. As this approach has a negative impact on TAC and utility costs, other approaches aim to minimise them [40]. The tradeoff between the targets is continuously changing from one study to the next, and it is unequivocal that the tradeoff has a significant effect on the structure of the designed HEN, along with the controllability and observability of the system.

The dynamical characteristics of the HENs have increasingly become the focus of attention, such as controllability [41, 48], observability [53], flexibility [18] or operability [8, 61]. The trends mentioned above highlight that it is more critical to automatically qualify the operability-related dynamical properties of HENs based on structural information. For this purpose, HENs can be transformed into the networks of state variables [31], streams and matches [37] or networks of state-space representations [4].

These promising representations could be beneficial to utilise the network science-based methodology to reveal the complexity of HENs in terms of their operability and to address the question concerning what kind of centrality measures can be applied to compare the complexity of HENs obtained by different design methodologies.

Section 4.2 presents the three different network-based representations of HENs to explore motifs that have a significant effect on structural controllability and observability. The utilised network-based measures are also introduced in this section. Section 4.3 presents how the proposed approach can be used in the systematic analysis of almost 50 HEN design problems and how the developed measures can be used to compare different design methodologies.

This chapter is based on the previous work of the authors [32], so most of the figures and tables are taken from this open-source publication.

4.2 Network-Based Complexity and Operability Analysis of HENs

The workflow of the proposed method is depicted in Fig. 4.1. The method handles several goal-oriented representations of HENs to analyse their specific properties. As this work focuses on graph-based approaches, approaches based on temperature-enthalpy, heat-content or temperature-interval diagrams [43] are not discussed. As our goal was to study dynamic structural characteristics, this section presents the three most relevant models that can be used for such a purpose. After the introduction of the network-based representations, the systematic analysis of the extracted network is presented (see Fig. 4.1). For this purpose, an Octave- and MATLAB-compatible toolbox was developed called NOCAD and will be introduced in Chap. 5.

Besides the widely applied process flow diagrams (PFDs), HENs of hot and cold streams are represented in three different ways, as illustrated in Fig. 4.2.

The first classical representation is the state-space (SS) approach (Fig. 4.2a), which consists of the distribution network (DN) and the superstructure operator [4]. The DN determines how units are located on the streams, while the superstructure operator shows the interactions between the streams. The SS-based network representation (Fig. 4.2b) can be defined as $G_{SS}(V_{SS}, E_{SS})$ based on the V_{SS} set of vertices that contain the inlet and outlet of the streams and the E_{SS} edges that present the connections between the heat exchangers.

The second classical representation shows how streams are matched according to the units, namely heat exchangers, utility heaters and utility coolers (Fig. 4.2c) [37]. The second SM-based network representation is almost the same as the classical

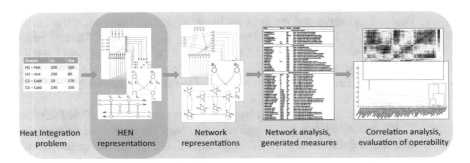

| Heat Integration problem | HEN representations | Network representations | Network analysis, generated measures | Correlation analysis, evaluation of operability |

Fig. 4.1 Workflow of the proposed methodology

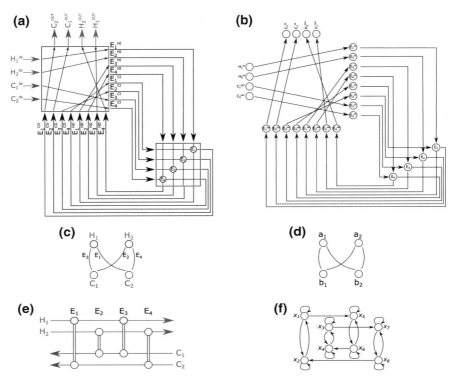

Fig. 4.2 Classical representations of HENs and their graph-based models. The representations are as follows: **a** state-space approach (SS); **b** network of state-space approach, where nodes are determined as the endpoints of the edges in the SS representation; **c** streams and matches (SM); **d** bipartite network representation of streams and matches; **e** Grid Diagram; and **f** network representation of state variables based on the differential equations describe the HEN

one (Fig. 4.2d), and it can be defined as $G_{SM}(V_{SM}, E_{SM})$, where the nodes ($V_{SM} = \{V_{hs}, V_{cs}, V_{hp}, V_{cw}\}$) denote the sets of hot streams (V_{hs}) and cold streams (V_{cs}), the high pressure steam (V_{hp}) and the cold water (V_{cw}). The edges ($E_{SM} = \{E_{he}, E_{uh}, E_{uc}\} \subseteq V_{SM} \times V_{SM}$) represent the heat exchangers ($E_{he} \subseteq V_{hs} \times V_{cs}$), the utility heaters ($E_{uh} \subseteq V_{hp} \times V_{cs}$) and the utility coolers ($E_{uc} \subseteq V_{hs} \times V_{cw}$). The two partitions of the nodes are the hot streams (Process and utility, $\{V_{hs} \cup V_{hp}\}$) and the cold streams (Process and utility, $\{V_{cs} \cup V_{cw}\}$). The weights of the nodes represent the heat capacity of the streams, while the weights of the edges represent the heat load on the units.

The third and most common classical representation of a HEN is the Grid Diagram (GD) [25, 35], where the streams are presented as they are connected through the heat exchangers and influenced by utility units (Fig. 4.2e). The dynamics of the units, i.e. the heat exchangers, utility coolers and utility heaters can be described by a differential-algebraic system of equations (DAE) as [54]:

$$\frac{dT_{ho}}{dt} = \frac{v_h}{V_h}(T_{hi} - T_{ho}) + \frac{UA}{c_{ph}\rho_h V_h}(T_{co} - T_{ho}), \qquad (4.1)$$

$$\frac{dT_{co}}{dt} = \frac{v_c}{V_c}(T_{ci} - T_{co}) + \frac{UA}{c_{pc}\rho_c V_c}(T_{ho} - T_{co}), \qquad (4.2)$$

where T_{hi}, T_{ci}, T_{ho} and T_{co} denote the temperatures of the hot input, cold input, hot output and cold output streams, respectively; v_h and v_c represent the flow rates of the cold and hot streams; V_h and V_c stand for the volumes of the hot and cold side tanks of the heat exchanger; U is the heat transfer coefficient; A denotes the heat transfer area of the heat exchanger; and c_p and ρ are the specific heat and density of the streams.

In the above representation, the state variables are the temperatures of the outlet streams of the heat exchanger $\mathbf{x}(t) = [T_{ho}, T_{co}]^T$, the temperature of the inlet streams are regarded as disturbances $\mathbf{d}(t) = [T_{hi}, T_{ci}]^T$ and $v_c(t)$ and $v_h(t)$ are time-varying parameters [54]. The dynamics of the utility units are similar. Utility coolers can be described by Eq. (4.1) by considering the temperature of the cold water stream T_{co} as a controlled input variable. Analogously, utility heaters are modeled by Eq. (4.2) where T_{ho} denotes the controlled inputs.

The resultant network can be described as a graph $G_{DAE} = (V_{DAE}, E_{DAE})$, where vertices represent the state variables $\mathbf{x}$, while the edges are derived from the structure matrix of the state-transition matrix $\mathbf{A}$ as if $\mathbf{A}_{ij} \neq 0$, then an edge from x_j to x_i exists. As a result of Eqs. (4.1) and (4.2), the simplest building blocks of HENs can be defined as the heat exchanger cells, the utility coolers and the utility heaters. Their Grid Diagram and state-space-based network representations can be seen in Fig. 4.3.

The SS-based network representation can be easily applied to HENs even when the streams are mixed and split. In the other two network representations, the handling of mixers and bypasses is less straightforward.

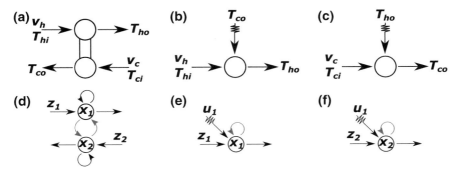

Fig. 4.3 The Grid Diagram and state-space network representations of: a heat exchanger cell (**a, d**); a utility cooler (**b, e**); and a utility heater (**c, f**). Red edges denote loops in the network representation, which create a strongly connected component in each elementary building block of HENs. In the case of the heat exchanger cell, the hot and cold sides of the exchanger belong to the same component even though they also have their own intrinsic dynamics, as in the case of utility units

The number of vertices and edges of the SM-based network representation directly represents the streams and units, respectively. Loops and paths can be identified more easily in this network representation than in any other representations. This property is essential when the goal is not the MER design but to minimise the number of units that can be readily determined by the equation $U_{un} = N_s + L - S$, where U_{un} denotes the number of units, N_s the number of streams ($|V_{GD}|$), L the number of independent loops and S the number of separate components in the network [37]. Loops and paths should be excluded from the designs, where the target is to minimise the number of units. The presence of splitters and mixers does not influence the property $U_{un} = N_s + L - S$ when Pinch Partitioning is excluded [58]. Furthermore, with this representation the degree of freedom of the heat loads can also be determined as $H - N_s + S$, where H denotes the number of heat exchangers ($|E_{he}|$) in the network.

The DAE-based network representation is suitable to analyse the operability of HENs.

As each motif in the DAE-based network representation possesses a self-loop (coloured in red in Fig. 4.3), i.e. the diagonal elements are non-zero values in the state-transition matrix, maximum matching selects these edges and leaves no unmatched node that can be a driver (or sensor) node. Since each HEN is constructed from these motifs, the path-finding method is required to determine satiable input and output configurations [30].

Table 4.1 Building elements of the studied network-based representations of HENs

Property	SS	SM	DAE
Node (V_i)	Connection of streams	Stream	State variable
Edge (E_i)	Stream	Match: a unit	Dynamical effect between two state variables
Node weight (n_i)	The nodes are not weighted	The heat load of the stream	The output temperature of the exchanger
Edge weight (w_i)	Edges are not weighted	Heat load exchanged between the streams	Dynamical effect between state variables determined by the state-transition matrix
Direction	The network is directed, edge directions are based on the direction of the streams [4]. The network is more transparent when—similarly to Grid Diagrams—the hot streams run from the left to the right and cold vice versa to follow the direction of Composite Curves	The network is undirected	The network is directed, the direction of edges is based on matrix **A** [38] that reflects the direction of the streams and the heat transfer
Loops	Loops can reflect superfluous units	Loops appear when there are more units than necessary	Loops create strongly connected components that can be controlled (or observed) by only one driver (or sensor) node
Maximum matching [38]	n/a	n/a	Unmatched nodes provide driver and sensor nodes
Relative degree [14]	n/a	n/a	Difficulty in operation
Number of components (S)	Components are the disconnected subnetworks in the HEN		

Table 4.2 Structural properties of the SS, SM and DAE-based network representations of HENs. The operator $|X|$ yields the cardinality of set X and the operator $\langle X \rangle$ yields the average of the values in X

Property	Calculation	SS	SM	DAE
Number of nodes	$N = |V|$	Starting- and endpoints of the streams	Number of streams (Process and utility)	Number of state variables necessary to describe the dynamics of HEN
Number of edges	$M = |E|$	Number of stream intervals	Number of units	Non-zero elements of the state-transition matrix
Degree	$k_i^{out} = \sum_{j \in V} A_{ij}$, $k_i^{in} = \sum_{j \in V} A_{ji}$, $k_i = k_i^{out} + k_i^{in}$	Fixed for the nodes, $k_i = 1$ for stream source and drains, $k_i = 2$ for endpoints of stream intervals and $k_i = 4$ for heat exchanger units	k_i is the number of units on stream	Fixed for the nodes, k_i moves from 2 to 4 in case of heat exchanger, and from 1 to 2 in case of utility units (loops are excluded)
Shortest path	$\ell_{i,j}$	The minimum of length of paths from node i to node j		
Distance matrix	$\mathcal{D}$	The distance matrix contains the shortest distance between node i and j, $\mathcal{D}_{i,j}$. The distance of node j is $\mathcal{D}_i = \sum_j \mathcal{D}_{ij}$		
Cycle rank [55]	L	The number of independent loops in the network		
Total walk count (TWC) [7]	$TWC(G) = \sum_{l=1}^{N-1} \sum_i \sum_j A^l$	The number of walks is proportional to the complexity, as the walks can represent the effect of an input signal in the system. More walks means more effect in the HEN		
Coefficient of network complexity (CNC) [28]	$CNC(G) = M^2/N$	Due to the fixed degree, the CNC is constrained	The proportion of units to streams	Due to the fixed degree, the CNC is constrained
Eccentricity [7]	$Ecc(i) = \max_{j \in V} (\ell_{i,j}) Ecc(G) = \max_{i \in V} (Ecc(i))$	Maximum length of heat loads	n/a	The relative degree of the system
Wiener index [15]	$W(G) = \sum_{(v,w) \in E} \ell_{v,w}$	Reflects the size of the HEN, and the complexity as long paths increase $W(G)$	In undirected case, $W(G)$ should be divided by two. Since the network is undirected, $W(G)$ approximates the size of the system	Longer geodesic paths represent higher order dynamics
A/D index [7]	$AD(G) = \langle k_i \rangle / \langle d_i \rangle$	Since the degree is fixed, the index approximates the closeness of the units	As the order of heat exchanger units is not given, the A/D index does not necessarily characterise heat loads or HEN	Since degree is fixed, the index approximates the closeness of units
Balaban-J index [6]	$BJ(G) = \frac{M}{L+1} \sum_{(v,w) \in E} \sqrt{\mathcal{D}_v \mathcal{D}_w}$	Balaban-J index can only be calculated to undirected graphs	It approximates the size of the HEN	Balaban-J index can only be calculated to undirected graphs
Connectivity index [15]	$CI(G) = \sum_{(v,w) \in E} 1/\sqrt{k_v k_w}$	Since the degree is fixed, the index cannot characterise the HEN	Shows how the streams are interconnected	Since the degree is fixed, the index cannot characterise the HEN

The analysis of the previously introduced network representations can highlight the structural properties of HENs. The developed network-based measures are summarised in Tables 4.1 and 4.2. For all HENs presented, the number of additional actuators and sensors required to manage the system with r_{max} was calculated by the set covering method (global optimum), the retrofit set covering method (using existing configurations) and the retrofit centrality-based method (using current configurations). In this analysis, the maximal-enabled relative degree was determined as $r_{max} = \lfloor d_{max}/4 \rfloor$, where d_{max} denotes the diameter of the network.

4.3   Studied Benchmark Problems

The well-known benchmark sets of Furman and Sahinidis (2004) [20, 33], Chen et al. (2015) [9, 10] and Grossmann (2017) [33] were used to study how the proposed measures can be used to evaluate HENs and compare different design methodologies of HENs. The benchmark problems and the applied methods are summarised in Table 4.3, while the notation of the utilised methods and their objective functions are shown in Table 4.4. These problem sets contain 48 problems and 23 methods, as well as 639 different HENs, of which 539 are unique. Fifty-three different measures, as summarised in Table 4.5, were generated for all the HENs.

4.3.1   Analysis of the 9sp-al1 Problem

The 9sp-al1 benchmark problem was studied in detail to demonstrate the applicability of the proposed methodology. As can be seen in Fig. 4.4, the solutions to the problem generated by the E, MER, EC and ECR methods significantly differed [34]. The first two solutions were based on the pinch point analysis to ensure Minimal Energy Requirement design, while the second two solutions were obtained by minimising energy and capital costs in the case of a new (EC) and retrofit (ECR) design.

The networks extracted from the HEN 9sp-al1-E solution (Fig. 4.4a) can be seen in Fig. 4.5. In DAE-based network representation (Fig. 4.5b), the driver and sensor nodes are denoted by green- and orange-coloured symbols, respectively, while in SM-based network representation, a critical path is denoted by the colour green (Fig. 4.5c).

The clusters (communities) in the DAE-based network representation were also analysed by community detection algorithms [3]. The clusters of the HEN 9sp-al1-MER can be seen in Fig. 4.6. Such clustering of the network can highlight useful information, e.g. it can show how the HEN is integrated and how the pinch isolates the clusters.

All measures introduced in Table 4.5 were calculated for the presented HENs, and the results of the analysis are presented in Table 4.6.

Table 4.3 Benchmark HENs and their optimisation methods that are used during the analysis. As the heuristic methods, CP, CRR, CSH, FLPR, GP, GSH, LFM, LHM, LHM-LP, LRR, SS, WFG and WFM were applied to each problem. This set of methods is denoted as HEU referring to heuristic methods. The methods are introduced in Table 4.4, where abbreviations are also presented

Problem	Hot streams	Cold streams	Methods	Source
Furman and Sahinidis (2004)				
4sp1	2	2	HEU, BB	[29]
6sp-cf1	3	3	HEU, RET	[12]
6sp-gg1	3	3	HEU	[22]
6sp1	3	3	HEU, BB	[29]
7sp-cm1	3	4	HEU	[13]
7sp-s1	6	1	HEU	[47]
7sp-torw1	4	3	HEU	[51]
7sp1	3	4	HEU, SYN	[40]
7sp2	3	3	HEU, SYN	[40]
7sp4	6	1	HEU	[16]
8sp-fs1	5	3	HEU, E	[19]
8sp1	4	4	HEU	[21]
9sp-al1	4	5	HEU, E, MER, EC, ECR, ST	[2]
9sp-has1	5	4	HEU	[23]
10sp-la1	4	5	HEU	[34]
10sp-ol1	4	6	HEU	[47]
10sp1	5	5	HEU, BB	[44]
12sp1	9	3	HEU	[21]
14sp1	7	7	HEU	[21]
15sp-tkm	9	6	HEU, E	[49]
20sp1	10	10	HEU	[21]
22sp-ph	11	11	HEU	[45]
22sp1	11	11	HEU	[5]
23sp1	11	12	HEU, DS	[42]
28sp-as1	16	12	HEU	[1]
37sp-yfyv	21	16	HEU, GASA	[59]
Chen et al. (2015a, b)				
balanced5	5	5	HEU	[9, 10]
balanced8	8	8	HEU	[9, 10]
balanced10	10	10	HEU	[9, 10]
balanced12	12	12	HEU	[9, 10]
balanced15	15	15	HEU	[9, 10]
unbalanced5	5	5	HEU	[9, 10]
unbalanced10	10	10	HEU	[9, 10]
unbalanced15	15	15	HEU	[9, 10]
unbalanced17	17	17	HEU	[9, 10]
unbalanced20	20	20	HEU	[9, 10]
Grossmann (2017)				
balanced12_random0	12	12	HEU	[33]
balanced12_random1	12	12	HEU	[33]

(continued)

Table 4.3 (continued)

Problem	Hot streams	Cold streams	Methods	Source
balanced12_random2	12	12	HEU	[33]
balanced15_random0	15	15	HEU	[33]
balanced15_random1	15	15	HEU	[33]
balanced15_random2	15	15	HEU	[33]
unbalanced17_random0	17	17	HEU	[33]
unbalanced17_random1	17	17	HEU	[33]
unbalanced17_random2	17	17	HEU	[33]
unbalanced20_random0	20	20	HEU	[33]
unbalanced20_random1	20	20	HEU	[33]
unbalanced20_random2	20	20	HEU	[33]

Table 4.4 Utilised methods and their objective functions

Name	Method	Objective function	Source
BB	Branch and bound	Minimises the heat exchanger and utility costs	[29]
CP	CPLEX-solved transportation model	Minimum number of matches	[33]
CRR	Covering Relaxation Rounding	Minimum number of matches	[33]
CSH	CPLEX-solved transshipment model	Minimum number of matches	[33]
DS	Decomposition Strategy	Minimises the number of exchangers	[42]
E	No method declared	Existing network, not optimised	
EC	Pinch methodology	Energy–Capital cost minimisation	[2, 34]
ECR	Pinch methodology	Energy–Capital Retrofit cost minimisation	[2, 34]
FLPR	Fractional LP Rounding	Minimum number of matches	[33]
GASA	Genetic Algorithm with Simulated Annealing	Minimises the annual cost	[59]
GP	Gurobi-solved transportation model	Minimum number of matches	[33]
GSH	Gurobi-solved transshipment model	Minimum number of matches	[33]
LFM	Largest Fraction Match	Minimum number of matches	[33]
LHM	Largest Heat Match Greedy	Minimum number of matches	[33]
LHM-LP	Largest Heat Match LP-based	Minimum number of matches	[33]
LRR	Lagrangian Relaxation Rounding	Minimum number of matches	[33]
MER	Pinch methodology	Maximum Energy Recovery, Minimum Energy Requirement	[2, 34]
RET	Structural modifications by categories	Minimises the cost of new exchangers, exchanger areas and piping	[12]
SS	Shortest Stream	Minimum number of matches	[33]
ST	Pinch methodology	Supertargeting	[2, 34]
SYN	Heuristic stage-by-stage structuring synthesis	Minimises the annual cost	[40]
WFG	Water Filling Greedy	Minimum number of matches	[33]
WFM	Water Filling MILP	Minimum number of matches	[33]

Table 4.5 Calculated measures and their short descriptions

Name	Measure	Network	Description
HEN-specific measures			
numOfHotStream		SM	Num. of the hot streams in the problem
numOfColdStream		SM	Num. of the hot streams in the problem
hasPinch		SS, DAE	True, if Pinch point is determined for the problem
numOfExchangers		SM	Num. of heat exchanger cells in the designed HEN
numOfUtilityHeater		SM	Num. of utility heater cells in the designed HEN
numOfUtilityCooler		SM	Num. of utility cooler cells in the designed HEN
numOfUnits	U_{un}	SM	Num. of heat exchanger units in the designed HEN
numOfMinUnits	U_{min}	SM	Num. of minimum units necessary for the problem
Dynamical properties			
numOfIndLoop	L	SM	Num. of independent loops
degOfFreedom		SM	Degree of freedom of heat loads
numOfCompDAE		DAE	Num. of subnetworks in the designed HEN
numOfCompSM	S	SM	Num. of components in the SM network
numOfUMCon		DAE	Num. of driver nodes granted by maximum matching
numOfDriver		DAE	Num. of actuator nodes necessary for controllability
numOfUMObs		DAE	Num. of sensor nodes granted by maximum matching
numOfSensor		DAE	Num. of sensor nodes necessary for observability
sluggishness	r	DAE	Relative degree of the HEN
targetDegree	r_{max}	DAE	Target relative degree (in our case: $\lfloor d_{max}/4 \rfloor$)
actCovSize		DAE	Num. of actuators determined by the set covering method
senCovSize		DAE	Num. of sensors determined by the set covering method
actCovRetSize		DAE	Num. of actuators determined by the retrofit set covering method
senCovRetSize		DAE	Num. of sensors determined by the retrofit set covering method
actNetMesSizeRet		DAE	Num. of actuators determined by the retrofit centrality-based method
senNetMesSizeRet		DAE	Num. of sensors determined by the retrofit centrality-based method
Network-based structural properties			
balabanJIndex	$BJ(G)$	SM	Balaban-J index of the SM network
numOfNodes.ofDAE	N	DAE	Num. of nodes in the DAE network
numOfNodes.ofSM	N	SM	Num. of nodes in the SM network
numOfNodes.ofSS	N	SS	Num. of nodes in the SS network
numOfEdges.ofDAE	M	DAE	Num. of edges in the DAE network
numOfEdges.ofSM	M	SM	Num. of edges in the SM network
numOfEdges.ofSS	M	SS	Num. of edges in the SS network
averageDegree.ofDAE	$\langle k_i \rangle$	DAE	Average degree in the DAE network
averageDegree.ofSM	$\langle k_i \rangle$	SM	Average degree in the SM network
averageDegree.ofSS	$\langle k_i \rangle$	SS	Average degree in the SS network
eccentricity.ofDAE	$Ecc(G)$	DAE	Graph eccentricity value of DAE network

(continued)

Table 4.5 (continued)

Name	Measure	Network	Description
eccentricity.ofSM	$Ecc(G)$	SM	Graph eccentricity value of SM network
eccentricity.ofSS	$Ecc(G)$	SS	Graph eccentricity value of SS network
totalWalkCount.ofDAE	$TWC(G)$	DAE	Num. of walks in the DAE network
totalWalkCount.ofSM	$TWC(G)$	SM	Num. of walks in the SM network
totalWalkCount.ofSS	$TWC(G)$	SS	Num. of walks in the SS network
coeffNetwCompl.ofDAE	$CNC(G)$	DAE	Coefficient of network complexity in the DAE network
coeffNetwCompl.ofSM	$CNC(G)$	SM	Coefficient of network complexity in the SM network
coeffNetwCompl.ofSS	$CNC(G)$	SS	Coefficient of network complexity in the SS network
adIndex.ofDAE	$AD(G)$	DAE	A/D index of the DAE network
adIndex.ofSM	$AD(G)$	SM	A/D index of the SM network
adIndex.ofSS	$AD(G)$	SS	A/D index of the SS network
wienerIndex.ofDAE	$W(G)$	DAE	Wiener index of the DAE network
wienerIndex.ofSM	$W(G)$	SM	Wiener index of the SM network
wienerIndex.ofSS	$W(G)$	SS	Wiener index of the SS network
connInd.ofDAE.inConn	$CJ(G)$	DAE	Connectivity index for in-degree in DAE network
connInd.ofDAE.outConn	$CJ(G)$	DAE	Connectivity index for out-degree in DAE network
connInd.ofSM	$CJ(G)$	SM	Connectivity index for SM network
connInd.ofSS.inConn	$CJ(G)$	SS	Connectivity index for in-degree in the SS network
connInd.ofSS.inConn	$CJ(G)$	SS	Connectivity index for out-degree in the SS network

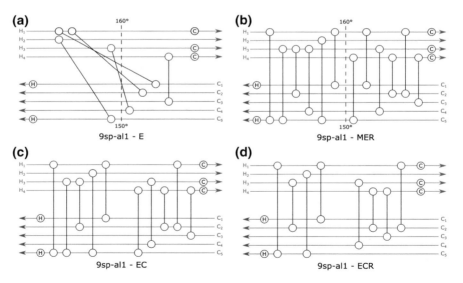

Fig. 4.4 Four designed HEN for the 9sp-al1 problem [34]: **a** existing network of an aromatic complexes in Europe; **b** maximum Energy Recovery design of the aromatic complexes; **c** grassroot (new) design with optimisation for the Energy–Capital tradeoff; and **d** retrofit design with optimisation for the Energy–Capital tradeoff

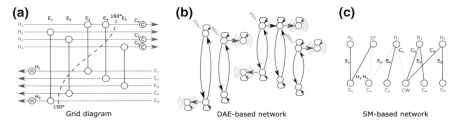

Fig. 4.5 Networks extracted from the HEN 9sp-al1-E: **a** Grid Diagram of the HEN seen in Fig. 4.4a; **b** DAE-based network representation, in which the green symbols represent the nodes where the input signal should be shared, and the orange symbols stand for the outputs to grant structural controllability and observability; and **c** SM-based network representation, in which green edges show paths between high-pressure steam and cold water, i.e. between utility units which should be broken in order to reach the minimal number of units. In this HEN, no loop appears

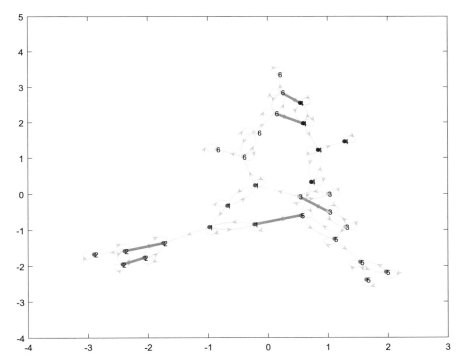

Fig. 4.6 Communities of the DAE-based network representation of HEN 9sp-al1-MER were also identified. The colours of the nodes and their labels denote the community IDs. The highlighted edges are the intercept of the pinch

As our analysis also aimed to identify structural parameters that have an impact on the dynamical properties of the HEN, firstly, the correlation between the proposed measures in these four examples was studied (see Fig. 4.7). The results are discussed in the following section together with the analysis of all 639 HENs.

Table 4.6 Calculated measures and their short descriptions

Name	9sp-all - E	9sp-all - MER	9sp-all - EC	9sp-all - ECR
HEN-specific measures				
numOfHotStream	4	4	4	4
numOfColdStream	5	5	5	5
hasPinch	0	1	0	0
numOfExchangers	5	12	10	8
numOfUtilityHeater	2	2	2	2
numOfUtilityCooler	3	3	3	3
numOfUnits	10	17	15	13
numOfMinUnits	10	10	10	10
Dynamical properties				
numOfIndLoop	0	7	5	3
degOfFreedom	−5	2	0	−2
numOfCompDAE	4	1	1	1
numOfCompSM	1	1	1	1
numOfUMCon	0	0	0	0
numOfDriver	4	1	1	1
numOfUMObs	0	0	0	0
numOfSensor	5	5	5	5
sluggishness	2	9	8	11
targetDegree	1	3	2	2
actCovSize	9	7	8	7
senCovSize	5	4	6	4
actCovRetSize	9	7	7	7
senCovRetSize	7	4	6	5
actNetMesSizeRet	9	8	10	8
senNetMesSizeRet	7	4	7	7
Network-based structural properties				
balabanJIndex	9.9212	4.3990	5.6234	5.2789
numOfNodesofDAE	15	29	25	21
numOfNodesofSM	11	11	11	11
numOfNodesofSS	72	107	97	87
numOfEdgesofDAE	31	73	61	49
numOfEdgesofSM	10	17	15	13
numOfEdgesofSS	71	113	101	89
averageDegreeofDAE	2.1333	3.0345	2.8800	2.6667
averageDegreeofSM	1.8182	3.0909	2.7273	2.3636
averageDegreeofSS	1.9722	2.1121	2.0825	2.0460
eccentricityofDAE	3	12	11	11
eccentricityofSM	7	5	4	5
eccentricityofSS	13	28	22	25
totalWalkCountofDAE	802,788	1.02127E+13	92,080,948,186	1,364,370,909
totalWalkCountofSM	34,258	2,114,490	2,114,490	410,032

(continued)

Table 4.6 (continued)

Name	9sp-al1 - E	9sp-al1 - MER	9sp-al1 - EC	9sp-al1 - ECR
totalWalkCountofSS	646	9138	3274	1977
coeffNetwComplofDAE	64.0667	183.7586	148.8400	114.3333
coeffNetwComplofSM	9.0909	26.2727	20.4545	15.3636
coeffNetwComplofSS	70.0139	119.3364	105.1649	91.0460
adIndexofDAE	0.8421	0.0522	0.0661	0.0787
adIndexofSM	0.1136	0.2636	0.2586	0.1857
adIndexofSS	0.0446	0.0067	0.0089	0.0107
wienerIndexofDAE	38	1687	1090	712
wienerIndexofSM	176	129	116	140
wienerIndexofSS	3,181	33,524	22,586	16,582
connIndofDAEIn	0.0154	0.0053	0.0065	0.0085
connIndofDAEOut	0.0147	0.0052	0.0063	0.0082
connIndofSS	0.0246	0.0082	0.0103	0.0143
connIndofSMIn	0.0131	0.0077	0.0087	0.0100
connIndofSMOut	0.0131	0.0077	0.0087	0.0100

4.3.2 Correlation Between Structural and Dynamical Properties

Figure 4.8 shows the similarity-based ordering of the 539 unique networks (on the rows) and the measures (on the columns). The colours of the map represent rankings of the networks based on the given measures. As can be seen, the map is clustered and on the right-hand side measures that are negatively correlated to the measures on the left-hand side are presented. This phenomenon originates from the calculation of the measures (some of these are calculated mostly as the reciprocal of the measures on the left-hand side). The similarities between the most important measures are visualised by a dendrogram in Fig. 4.9.

Based on the analysis of the results, the following conclusions can be made:

Firstly, it is visible that the studied HENs were clustered and the members of such clusters exhibited similar dynamical properties. To confirm this, the Spearman distance for each network based on the measures was generated and visualised on a heat map, as can be seen in Fig. 4.10. On the heat map, three groups can be easily distinguished, and these groups consisted mostly of problems similar in size.

Secondly, the number of unmatched nodes that were generated by maximum matching was equal to zero for all HENs. This was caused by the SCCs and the intrinsic dynamics introduced in Fig. 4.3. Thus, additional driver and sensor nodes should be determined by methods such as path-finding [30].

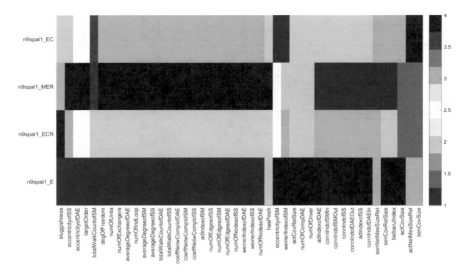

Fig. 4.7 Correlations between the proposed network measures according to four HENs of the 9sp-all problem. The rows and columns were ordered based on similarities calculated based on the Spearman's rank correlation. The colours of the map represent the rank values. It can be determined, that a subset of the measures (on the right-hand side) exhibits a negative correlation with other measures. Based on this information, it is deducible that different design methods of HENs yield different results as reflected in the topology of the designed HEN

Thirdly, the number of the additional driver and sensor nodes correlated with the number of components in DAE-based representation, i.e., the number of subnetworks in the HEN. Thus, more interconnected streams demanded fewer driver and sensor nodes.

Fourthly, a smaller number of driver and sensor nodes resulted in an increased relative degree, thus more complex operability.

Fifthly, the number of independent loops in the SM-based network representation correlated with the number of units in the HEN, and the average degree in SS- and SM-based network representations. The number of loops also correlated negatively with connectivity indexes. As the number of independent loops resulted in enhanced heat recovery, it attracted an increase in the number of heat exchangers too. Thus, a larger average degree meant that energy integration was greater, and more heat exchangers were utilised.

Finally, the comparison of the proposed methods showed that the results significantly varied from problem to problem. Only one correlation could be determined: the CRR, LHM and LRR algorithms solved the problem using more exchangers than other methods.

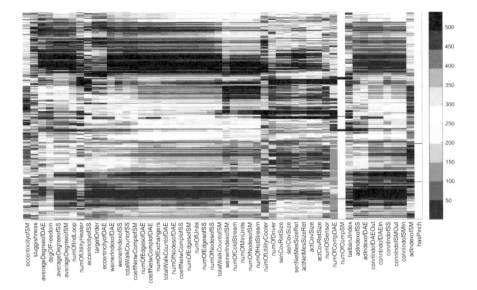

Fig. 4.8 Similarity-based ordering of the 539 HENs (rows) and developed measures (columns). The networks are ranked according to each measure, the similarities were calculated by Spearman's rank correlation and the colours of the map represent the rank values. The negative correlation between the measures on the right- and left-hand side is still detected

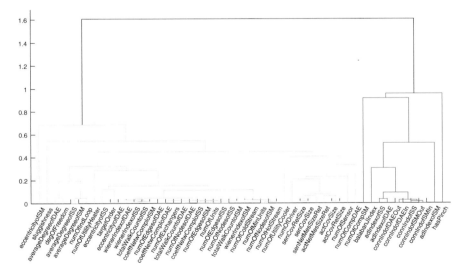

Fig. 4.9 Correlations between the measures. Dendrogram presenting the distances between the measures for all unique HENs

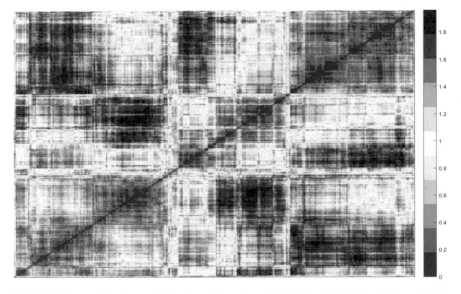

Fig. 4.10 Spearman's rank correlation-based clustering of 539 unique HENs. The clusters of HENs can be easily detected. The method ordered the networks according to their similarities which were mainly determined by the complexity and the applied design methodology

4.4 Discussion

In this section, a network science-based analytical approach was proposed for the structural analysis of heat exchanger networks.

The proposed methodology utilises three different network-based representations that highlight various aspects of the dynamics of the HEN. The extracted networks were analysed by the toolbox of network science to create structural operability and complexity measures.

The analysis of more than 600 HENs confirmed that the proposed approach can be efficiently applied to the fast screening of HENs based on their structural properties.

A significant result of the analysis is that the popular maximum matching-based method used to determine the sensor and driver nodes in dynamical systems was not suitable for heat exchanger networks due to the intrinsic dynamics. It was highlighted that, with the more interconnected HENs, fewer actuators and sensors are needed to ensure structural controllability and observability, which results in a higher relative degree and the increased difficulty of the operability.

It was also highlighted that the degree-based structural measures can refer to the level of Heat Integration. The methodology was suitable for classifying the different methods that grant the HEN, as some techniques tended to determine more units in the network than others.

Although the proposed methodology has been introduced for the analysis of HENs, it can be applied to a broader class of dynamical systems. The problem of the analysis of HENs was selected since it has been thoroughly researched using a diverse set of benchmark examples that can be represented as networks. The analysis of other problem sets would also be interesting. If the solutions of a design problem can be represented in different topologies, like in the case of HENs, it would be interesting to study what kind of useful information can be revealed by the structural analyses of the results. The toolbox which can support this analysis will be presented in the next chapter.

References

1. Ahmad, S., Smith, R.: Targets and design for minimum number of shells in heat exchanger networks. Chem. Eng. Res. Des. **67**(5), 481–494 (1989)
2. Ahmad, S., Linnhoff, B.: Supertargeting: different process structures for different economics. J. Energy Resour. Technol. **111**(3), 131–136 (1989)
3. Arenas, A., Fernandez, A., Gomez, S.: Analysis of the structure of complex networks at different resolution levels. New J. Phys. **10**(5), 053039 (2008)
4. Bagajewicz, M.J., Manousiouthakis, V.: Mass/heat-exchange network representation of distillation networks. AIChE J **38**(11), 1769–1800 (1992)
5. Bagajewicz, M.J., Pham, R., Manousiouthakis, V.: On the state space approach to mass/heat exchanger network design. Chem. Eng. Sci. **53**(14), 2595–2621 (1998)
6. Balaban, A.T.: Highly discriminating distance-based topological index. Chem. Phys. Lett. **89**(5), 399–404 (1982)
7. Bonchev, D., Buck, G.A.: Quantitative measures of network complexity. Complexity in Chemistry, Biology, and Ecology, pp. 191–235. Springer, Springer Science & Business Media, USA (2005)
8. Calandranis, J., Stephanopoulos, G.: Structural operability analysis of heat exchanger networks. Chem. Eng. Res. Des. (Icheme), **64**(5), 347–364 (1986)
9. Chen, Y., Grossmann, I.E., Miller, D.C.: Computational strategies for large-scale milp transshipment models for heat exchanger network synthesis. Comput. Chem. Eng. **82**, 68–83 (2015)
10. Chen, Y., Grossmann, I.E., Miller, D.C.: Large-scale milp transshipment models for heat exchanger network synthesis. Available from CyberInfrastructure for MINLP [www.minlp.org, a collaboration of Carnegie Mellon University and IBM Research] at: www.minlp.org/library/problem/index.php (2015)
11. Chrissis, M.B., Konrad, M., Shrum, S.: CMMI Guidlines for Process Integration and Product Improvement. Addison-Wesley Longman Publishing Co., Inc., USA (2003)
12. Ciric, A.R., Floudas, C.A.: A retrofit approach for heat exchanger networks. Comput. Chem. Eng. **13**(6), 703–715 (1989)
13. Colberg, R.D., Morari, M.: Area and capital cost targets for heat exchanger network synthesis with constrained matches and unequal heat transfer coefficients. Comput. Chem. Eng. **14**(1), 1–22 (1990)
14. Daoutidis, P., Kravaris, C.: Structural evaluation of control configurations for multivariable nonlinear processes. Chem. Eng. Sci. **47**(5), 1091–1107 (1992)
15. Dehmer, M., Kraus, V., Emmert-Streib, F., Pickl, S.: Quantitative Graph Theory. CRC Press, USA (2014)
16. Dolan, W.B., Cummings, P.T., Le Van, M.D.: Algorithmic efficiency of simulated annealing for heat exchanger network design. Comput. Chem. Eng. **14**(10), 1039–1050 (1990)
17. Düştegör, D., Frisk, E., Cocquempot, V., Krysander, M., Staroswiecki, M.: Structural analysis of fault isolability in the damadics benchmark. Control Eng. Pract. **14**(6), 597–608 (2006)

18. Escobar, M., Trierweiler, J.O., Grossmann, I.E.: Simultaneous synthesis of heat exchanger networks with operability considerations: flexibility and controllability. Comput. Chem. Eng. **55**, 158–180 (2013)
19. Farhanieh, B., Sunden, B.: Analysis of an existing heat exchanger network and effects of heat pump installations. Heat Recover. Syst. CHP **10**(3), 285–296 (1990)
20. Furman, K.C., Sahinidis, N.V.: Approximation algorithms for the minimum number of matches problem in heat exchanger network synthesis. Ind. Eng. Chem. Res. **43**(14), 3554–3565 (2004)
21. Grossmann, I.E., Sargent, R.W.H.: Optimum design of heat exchanger networks. Comput. Chem. Eng. **2**(1), 1–7 (1978)
22. Gundersen, T., Grossmann, I.E.: Improved optimization strategies for automated heat exchanger network synthesis through physical insights. Comput. Chem. Eng. **14**(9), 925–944 (1990)
23. Hall, S.G., Ahmad, S., Smith, R.: Capital cost targets for heat exchanger networks comprising mixed materials of construction, pressure ratings and exchanger types. Comput. Chem. Eng. **14**(3), 319–335 (1990)
24. Jamaluddin, K., Wan Alwi, S.R., Manan, Z.A., Klemes, JJ.: Pinch analysis methodology for trigeneration with energy storage system design. Chem. Eng. Tran. **70**, 1885–1890 (2018)
25. Kemp, I.C.: Pinch analysis and process integration: a user guide on process integration for the efficient use of energy. Elsevier, USA (2011)
26. Klemes, J.J.: Handbook of process integration (PI): minimisation of energy and water use, waste and emissions. Elsevier, UK (2013)
27. Klemes, J.J., Varbanov, P.S., Walmsley, T.G., Jia, X.: New directions in the implementation of pinch methodology (pm). Renew. Sustain. Energy Rev. **98**, 439–468 (2018)
28. Latva-Koivisto, A.M.: Finding a complexity measure for business process models. Helsinki University of Technology, Systems Analysis Laboratory (2001)
29. Lee, K.-F., Masso, A.H., Rudd, D.F.: Branch and bound synthesis of integrated process designs. Ind. Eng. Chem. Fundam. **9**(1), 48–58 (1970)
30. Leitold, D., Vathy-Fogarassy, Á., Abonyi, J.: Controllability and observability in complex networks-the effect of connection types. Sci. Rep. **7**, 151 (2017)
31. Leitold, D., Vathy-Fogarassy, A., Abonyi, J.: Network distance-based simulated annealing and fuzzy clustering for sensor placement ensuring observability and minimal relative degree. Sensors **18**(9), 3096 (2018)
32. Leitold, Dá., Vathy-Fogarassy, Á., Abonyi, J.: Evaluation of the complexity, controllability and observability of heat exchanger networks based on structural analysis of network representations. Energies **12**(3), 513 (2019)
33. Letsios, D., Kouyialis, G., Misener, R.: Heuristics with performance guarantees for the minimum number of matches problem in heat recovery network design. Comput. Chem. Eng. **113**, 57–85 (2018)
34. Linnhoff, B., Ahmad, S.: Supertargeting: optimum synthesis of energy management systems. J. Energy Resour. Technol. **111**(3), 121–130 (1989)
35. Linnhoff, B., Flower, J.R.: Synthesis of heat exchanger networks: I. systematic generation of energy optimal networks. AIChE J **24**(4), 633–642 (1978)
36. Linnhoff, B., Hindmarsh, E.: The pinch design method for heat exchanger networks. Chem. Eng. Sci. **38**(5), 745–763 (1983)
37. Linnhoff, B., Mason, D.R., Wardle, I.: Understanding heat exchanger networks. Comput. Chem. Eng. **3**(1–4), 295–302 (1979)
38. Liu, Y.-Y., Slotine, J.-J., Barabási, A.-L.: Controllability of complex networks. Nature **473**(7346), 167 (2011)
39. Liu, Y.-Y., Slotine, J.-J., Barabási, A.-L.: Observability of complex systems. Proc. Natl. Acad. Sci. **110**(7), 2460–2465 (2013)
40. Masso, A.H., Rudd, D.F.: The synthesis of system designs. ii. heuristic structuring. AIChE J **15**(1), 10–17 (1969)
41. Miranda, C.B., Costa Costa, C.B.B., Andrade, C.M.G., Ravagnani, M.A.S.S.: Controllability and resiliency analysis in heat exchanger networks. Chem. Eng. Trans. **61**, 1609–1614 (2017)

42. Mocsny, D., Govind, R.: Decomposition strategy for the synthesis of minimum-unit heat exchanger networks. AIChE J. **30**(5), 853–856 (1984)
43. Nishida, N., Stephanopoulos, G., Westerberg, A.W.: A review of process synthesis. AIChE J. **27**(3), 321–351 (1981)
44. Pho, T.K., Lapidus, L.: Topics in computer-aided design: Part ii. synthesis of optimal heat exchanger networks by tree searching algorithms. AIChE J. **19**(6), 1182–1189 (1973)
45. Polley, G.T., Heggs, P.J.: Don't let the pinch pinch you. Chem. Eng. Prog. **95**(12), 27–36 (1999)
46. Saboo, A.K., Morari, M., Woodcock, D.C.: Design of resilient processing plants-viii. a resilience index for heat exchanger networks. Chem. Eng. Sci. **40**(8), 1553–1565 (1985)
47. Shenoy, U.V.: Heat Exchanger Network Synthesis: Process Optimization by Energy and Resource Analysis. Gulf Professional Publishing, USA (1995)
48. Svensson, E., Eriksson, K., Bengtsson, F., Wik, T.: Design of heat exchanger networks with good controllability. Technical Report, CIT Industriell Energi AB, Sweden (2018)
49. Tantimuratha, L., Kokossis, A.C., Müller, F.U.: The heat exchanger network design as a paradigm of technology integration. Appl. Therm. Eng. **20**(15–16), 1589–1605 (2000)
50. Trivedi, K.K., O'Neill, B.K., Roach, J.R.: Synthesis of heat exchanger networks featuring multiple pinch points. Comput. Chem. Eng. **13**(3), 291–294 (1989)
51. Trivedi, K.K., O'Neill, B.K., Roach, J.R., Wood, R.M.: Systematic energy relaxation in mer heat exchanger networks. Comput. Chem. Eng. **14**(6), 601–611 (1990)
52. Varbanov, P.S., Walmsley, T.G., Klemes, J.J., Wang, Y., Jia, X.-X.: Footprint reduction strategy for industrial site operation. Chem. Eng. Trans. **67**, 607–612 (2018)
53. Varga, E.I., Hangos, K.M.: The effect of the heat exchanger network topology on the network control properties. Control Eng. Pract. **1**(2), 375–380 (1993)
54. Varga, E.I, Hangos, K.M., Szigeti, F.: Controllability and observability of heat exchanger networks in the time-varying parameter case. Control Eng. Pract. **3**(10), 1409–1419 (1995)
55. Volkmann, L.: Estimations for the number of cycles in a graph. Period. Math. Hung. **33**(2), 153–161 (1996)
56. Westphalen, D.L., Young, B.R., Svrcek, W.Y.: A controllability index for heat exchanger networks. Ind. Eng. Chem. Res. **42**(20), 4659–4667 (2003)
57. Wood, R.M., Suaysompol, K., O'Neill, B.K., Roach, J.R., Trivedi, K.K.: A new option for heat exchanger network design. Chem. Eng. Prog. **87**(9), 38–43 (1991)
58. Wood, R.M., Wilcox, R.J., Grossmann, I.E.: A note on the minimum number of units for heat exchanger network synthesis. Chem. Eng. Commun. **39**(1-6), 371–380 (1985)
59. Yu, H., Fang, H., Yao, P., Yuan, Y.: A combined genetic algorithm/simulated annealing algorithm for large scale system energy integration. Comput. Chem. Eng. **24**(8), 2023–2035 (2000)
60. Zafiriou, E.: The Integration of Process Design and Control. Pergamon, UK (1994)
61. Zhelev, T.K., Varbanov, P.S., Seikova, I.: Hen's operability analysis for better process integrated retrofit. Hung. J. Ind. Chem. **26**(2), 81–88 (1998)

Chapter 5
The NOCAD MATLAB/Octave Toolbox Developed for the Analysis of Dynamical Systems

Abstract In this chapter, the functionality of the proposed Octave- and MATLAB-compatible network-based observability and controllability analysis of dynamical systems (NOCAD) toolbox is introduced. This toolbox provides a set of methods, which enables the structural controllability and observability analysis of dynamical systems presented in the earlier chapters. The implemented functions are also demonstrated.

Keywords Dynamical systems · Complex networks · Controllability and observability analysis · Robustness · MATLAB toolbox · Octave toolbox

5.1 Existing Tools for Network Analysis

The network science-based analysis of dynamical systems has spread rapidly as it provides simple and efficient tools to analyse the structural controllability of any linear or linearised system [13]. Although considerable research has utilised the method [12], a flexible software tool which may be used to support the research in this field is yet to be designed. Parallel research has resulted in a collection of applications, toolboxes, plug-ins and scripts that analyse and determine several structural properties of genes, protein–protein interaction or even social or urban networks. Most of these applications only analyse the structural properties of static networks and just a handful of them utilise these structural properties to draw conclusions concerning the dynamics of the system investigated. As our toolbox belongs to the second group, in the following section, the available applications and programs of this group are elaborated.

This chapter is based on the previous work of the authors [11], so most of the figures and tables are taken from this open-source publication.

A brief summary of the available tools with expanded functionalities is given in Table 5.1. Applications or software packages implemented in Python and capable of analysing the controllability and observability of dynamical systems are: graph-control [4] and WDNfinder [5]. The advantage of Python-based development lies in its widespread use and the countless methods and packages implemented in this lan-

D. Leitold et al., *Network-Based Analysis of Dynamical Systems*,
SpringerBriefs in Computer Science,
https://doi.org/10.1007/978-3-030-36472-4_5

Table 5.1 Toolboxes that implement some functions for dynamical analysis of complex systems based on their structural analysis

Software	Language	Applied on	GUI	Refs.	Last updated
netctrl	C++	General networks	No	[14]	January 8, 2015
CONTEST	MATLAB	General networks	No	[17]	February 2009
CytoCtrlAnalyser	Java	Biomolecular networks	Yes	[18]	May 25, 2017
Graph-control	Python	General networks	No	[4]	December 16, 2015
WDNfinder	Python	Biological networks	No	[5]	June 24, 2018
enaR	R	Ecological networks	No	[2]	May 18, 2018

guage, including the tools developed for network analysis [19]. Although in Python the focus is on developing a broad software package for complex system analysis, this is yet to be fulfilled and all of the available solutions have limitations. The graph-control toolbox only analyses the impact of network topology on the number of inputs and implements the fast matching algorithm [6]. Even though WDNFinder only determines the minimum driver node set (MDS) and classifies nodes based on MDS, it is incapable of facilitating extended analysis.

Additionally, the CytoCtrlAnalyser [18] plug-in for Cytoscape [16] has been developed, which was implemented in Java and offers graphical interfaces for users as well. It evaluates control centrality, control capacity and classifies nodes for biomolecular networks. Furthermore, the Ecological Network Analysis with R software package (enaR) provides some dynamical analysis functions and can generate models to analyse ecological networks in the R environment [2]. As can be seen, both software packages deal with special kinds of networks. The netctrl program can determine the driver nodes and switchboard dynamics model for any complex network [14]. CONTEST is a MATLAB toolbox which can analyse the dynamics of complex systems, but these dynamics do not cover the structural controllability and observability properties [17] of the analysed system. Although the presented software packages ensure the design of a controllable and observable system, they do not provide the opportunity to analyse the designed system exhaustively. These functions are helpful in terms of supporting the work of experts, but are insufficient for the sophisticated analysis of systems.

The contribution of this chapter is to provide a novel toolbox, NOCAD [1], for the comprehensive analysis of linear or linearised dynamical systems based on the approach of network science. In the following section, the implemented functions and measurements are presented through examples of their application.

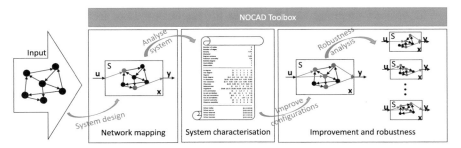

Fig. 5.1 Workflow of the utilisation of the NOCAD toolbox. The network mapping module provides two methods to create a dynamical system based on the topology of the state variables. The system characterisation module generates more than 49 measures to analyse, classify and characterise the developed system. The improvement and robustness module offers five algorithms to improve the system with additional inputs (observers) as well as outputs (controllers), and can analyse the robustness of the designed system

5.2   The Workflow Related to the Application of the Toolbox

With the help of the presented Ocatave- and MATLAB-compatible toolbox, experts can create, analyse and improve any type of dynamical systems. As the structure of the dynamical systems is generally represented by their adjacency matrix and linear dynamical systems can be described by the state-space model that contains the state-transition, input, output and feedthrough matrices, the Octave/MATLAB programming language is a perfect environment to handle these matrices and provide comprehensive functionalities based on them. With the use of NOCAD [1], experts and researchers can effectively determine the input and output matrices of state-space models, calculate system-specific qualitative measurements (e.g. diameter, relative degree, control centrality and robustness of the system, etc.) and improve the system to satisfy the relative degree-based requirements. The workflow of the toolbox can be seen in Fig. 5.1.

The functions of the toolbox can be performed step-by-step given its modular structure. Each module has a specific task and one function from each module calls the others. A system can be analysed by calling the main functions from the modules. The advantage of this structure is its modularity as each module can be expanded easily and further modules are also implemented in a simple way. A list of their functions and dependencies on each other is presented in the manual.

5.3   Structure of the Toolbox

According to the aforementioned approach, the implemented functions of the toolbox were divided into three modules as follows: (1) network mapping module, (2) system characterisation module and (3) improvements and robustness module.

The *network mapping module* creates a dynamical system from a given network structure, i.e. the necessary matrices of the state-space model are generated for the topology in such a way, that the created system is structurally controllable and structurally observable. The determination of the input and output matrices can be achieved by the path-finding and signal sharing methods (see more in Appendix A.1), which modify the result of the maximum matching algorithm.

The *system characterisation module* performs the calculation of 49 numerical measures to qualify the dynamical system based on its structure. The implemented measures, on the one hand, are well-known static measures (e.g. the number of nodes and edges, closeness and betweenness centralities), and, on the other hand measures that characterise the dynamics of the system (e.g. structural controllability, observability, control centrality and relative degree). This module can also be used for the purpose of simple network analysis.

The *improvement and robustness module* integrates two main functions. On the one hand, it enables the input and output configurations of the system to be extended in such a way that the relative degree of the modified system does not exceed the initially defined threshold (see more in Chap. 3). For this purpose, this module implements five methods, namely the set covering -based grassroot and retrofit methods [10], the centrality measures-based method [10], the modified Clustering Large Applications based on Simulated Annealing algorithm (mCLASA)and the Geodesic Distance-based Fuzzy c-Medoid Clustering with Simulated Annealing algorithm (GDFCMSA) [9, 10]. On the other hand, this module allows users to examine the robustness of the extended configurations by removing nodes from the network representation and by checking the structural controllability and structural observability of the damaged system.

The implemented methods are introduced in detail in Chap. 3 and the manual of the NOCAD toolbox [1]. In order to use the NOCAD toolbox [1], installation of Octave or MATLAB is required. Then the directories of the toolbox must be copied into the working directory, or the directories of the toolbox must be added to the paths. The functions were implemented in Octave 5.1.0 and MATLAB R2016a on a Windows 64-bit system. On other operating systems, or with other Octave or MATLAB versions, proper operation is not guaranteed. Our toolbox is independent of other MathWorks toolboxes, it uses only the octave-networks-toolbox [3] and the greedy set covering implementation [8].

The URL link to where the source code can be downloaded under the GNU General Public License v3.0: https://github.com/abonyilab/NOCAD and from Zenodo [1].

5.4   Use Cases

In this section, the main functionalities of the NOCAD toolbox [1] is presented through examples of use cases. Although many biological networks are available from public databases, due to their complex nature, they are unsuitable for such a

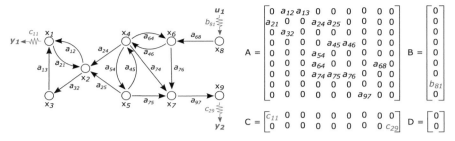

Fig. 5.2 An example network with determined input (blue) and output (red) configurations according to the path-finding method

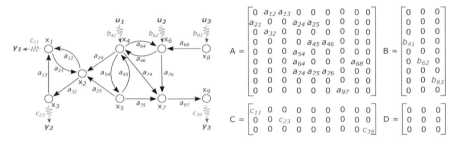

Fig. 5.3 The topology and associated configuration of the system for demonstration of the system characterisation module of the NOCAD toolbox

simple illustration. Therefore, the services of the NOCAD toolbox are presented on a simple artificial network.

The first step in each workflow is to create a state-space model from the adjacency matrix that presents the structural description of the system. This can be achieved by the use of path-finding and signal sharing methods implemented in the first module. Both methods are modified versions of the maximum matching algorithm. An example of the application of the path-finding method for the creation of a state-space model from the adjacency matrix ($\mathbf{A}$) is shown in Fig. 5.2.

As the configuration above is not complex enough to demonstrate the functions of the second module, a more complex configuration of the input and output nodes is used. The sample input and output configurations can be seen in Fig. 5.3, where the input and the output nodes are denoted by blue and red, respectively.

The system presented in Fig. 5.3 consists of 9 state variables and 15 directed connections between them. Quality measures calculated by the system characterisation module of the NOCAD toolbox can be seen in Figs. 5.4, 5.5 and 5.6.

In Fig. 5.4, measures qualifying the whole system with one value are presented. The density shows that the number of edges is almost a fifth of the possible maximum, and the diameter of the system (i.e. the longest shortest path in the network that presents its structure) is 4. The degree variance is 2.67, while the Freeman's centrality is 0.43. The relative degree of the system is also 4. The Pearson coefficient shows

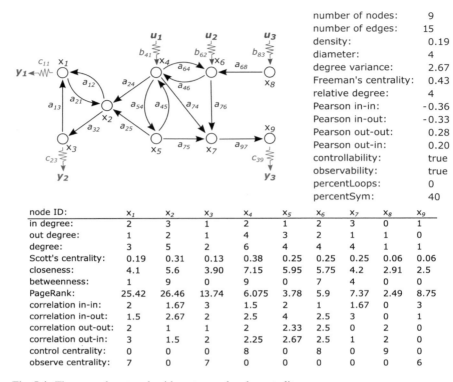

number of nodes:	9
number of edges:	15
density:	0.19
diameter:	4
degree variance:	2.67
Freeman's centrality:	0.43
relative degree:	4
Pearson in-in:	-0.36
Pearson in-out:	-0.33
Pearson out-out:	0.28
Pearson out-in:	0.20
controllability:	true
observability:	true
percentLoops:	0
percentSym:	40

node ID:	x_1	x_2	x_3	x_4	x_5	x_6	x_7	x_8	x_9
in degree:	2	3	1	2	1	2	3	0	1
out degree:	1	2	1	4	3	2	1	1	0
degree:	3	5	2	6	4	4	4	1	1
Scott's centrality:	0.19	0.31	0.13	0.38	0.25	0.25	0.25	0.06	0.06
closeness:	4.1	5.6	3.90	7.15	5.95	5.75	4.2	2.91	2.5
betweenness:	1	9	0	9	0	7	4	0	0
PageRank:	25.42	26.46	13.74	6.075	3.78	5.9	7.37	2.49	8.75
correlation in-in:	2	1.67	3	1.5	2	1	1.67	0	3
correlation in-out:	1.5	2.67	2	2.5	4	2.5	3	0	1
correlation out-out:	2	1	1	2	2.33	2.5	0	2	0
correlation out-in:	3	1.5	2	2.25	2.67	2.5	1	2	0
control centrality:	0	0	0	8	0	8	0	9	0
observe centrality:	7	0	7	0	0	0	0	0	6

Fig. 5.4 The example network with system and node centrality measures

that the in–in and in–out correlations are assortative in nature, while out–out and out–in correlations are likely to be disassortative. The system is controllable and observable. As no loop is present in the network, the percentage of loops relative to edges is 0%. As there are 6 edges that have symmetric edge pairs and the number of connections is 15, the percentage of the symmetric edge pairs relative to the edges is 40%.

Node centrality measures assigned to the state variables of the system are also presented in Fig. 5.4. One of the most important values is the highest degree of the nodes, which belongs to state variable x_4. As Scott's centrality is a normalised degree, the most important node is once again x_4. The closeness of node x_i is calculated as the ratio of the number of nodes reachable from x_i to the sum of their distances from x_i. The higher value indicates the more central position of the node, and, once again, node x_4 is the most central element. The betweenness centrality shows how many shortest paths intercept the given node. If a node has a high value, then it is a critical node in the structure. The highest value belongs to nodes x_2 and x_4. The PageRank assigns a percentage value for each node, based on their centrality roles if Markov-chains are modelled. The measure referred to as correlation shows the proportion of the number of edges of neighbours' and the number of neighbours. This information

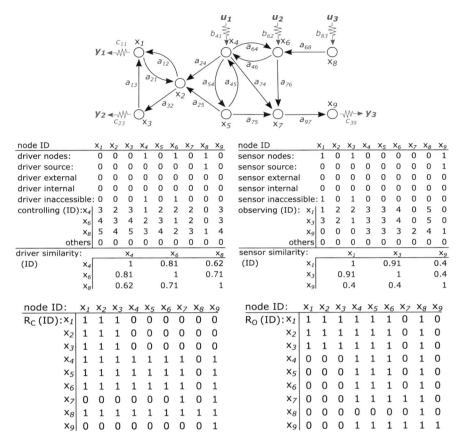

node ID	x_1	x_2	x_3	x_4	x_5	x_6	x_7	x_8	x_9
driver nodes:	0	0	0	1	0	1	0	1	0
driver source:	0	0	0	0	0	0	0	1	0
driver external	0	0	0	0	0	0	0	0	0
driver internal	0	0	0	0	0	0	0	0	0
driver inaccessible:	0	0	0	1	0	1	0	0	0
controlling (ID):x_4	3	2	3	1	2	2	2	0	3
x_6	4	3	4	2	3	1	2	0	3
x_8	5	4	5	3	4	2	3	1	4
others	0	0	0	0	0	0	0	0	0

node ID	x_1	x_2	x_3	x_4	x_5	x_6	x_7	x_8	x_9
sensor nodes:	1	0	1	0	0	0	0	0	1
sensor source:	0	0	0	0	0	0	0	0	1
sensor external	0	0	0	0	0	0	0	0	0
sensor internal	0	0	0	0	0	0	0	0	0
sensor inaccessible:	1	0	1	0	0	0	0	0	0
observing (ID): x_1	1	2	2	3	3	4	0	5	0
x_3	3	2	1	3	3	4	0	5	0
x_9	0	0	0	3	3	3	2	4	1
others	0	0	0	0	0	0	0	0	0

driver similarity:(ID)		x_4	x_6	x_8
	x_4	1	0.81	0.62
	x_6	0.81	1	0.71
	x_8	0.62	0.71	1

sensor similarity:(ID)		x_1	x_3	x_9
	x_1	1	0.91	0.4
	x_3	0.91	1	0.4
	x_9	0.4	0.4	1

node ID:	x_1	x_2	x_3	x_4	x_5	x_6	x_7	x_8	x_9
R_C (ID):x_1	1	1	1	0	0	0	0	0	0
x_2	1	1	1	0	0	0	0	0	0
x_3	1	1	1	0	0	0	0	0	0
x_4	1	1	1	1	1	1	1	0	1
x_5	1	1	1	1	1	1	1	0	1
x_6	1	1	1	1	1	1	1	0	1
x_7	0	0	0	0	0	0	1	0	1
x_8	1	1	1	1	1	1	1	1	1
x_9	0	0	0	0	0	0	0	0	1

node ID:	x_1	x_2	x_3	x_4	x_5	x_6	x_7	x_8	x_9
R_O (ID):x_1	1	1	1	1	1	1	0	1	0
x_2	1	1	1	1	1	1	0	1	0
x_3	1	1	1	1	1	1	0	1	0
x_4	0	0	0	1	1	1	0	1	0
x_5	0	0	0	1	1	1	0	1	0
x_6	0	0	0	1	1	1	0	1	0
x_7	0	0	0	1	1	1	1	1	0
x_8	0	0	0	0	0	0	0	1	0
x_9	0	0	0	1	1	1	1	1	1

Fig. 5.5 Measures of node clustering and the representation of the topology of the network

is useful when determining the assortativity of the system. The control centrality and observe centrality measures determine how many state variables can be influenced or observed by the nodes.

In Fig. 5.5, the first vectors (referred to as driver and sensor nodes) show the driver and sensor nodes as logical vectors. The following four vectors classify these nodes as source, external, internal and inaccessible driver and sensor nodes. These types of nodes are introduced in [15] in detail. In the next section of the figure, the controlling and observing matrices are presented. Generally, these matrices are sparse matrices, as only the columns of drivers and sensors contain non-zero values. In Fig. 5.5, we converted them into row vectors for their appropriate visualisation. The values show the number of derivations necessary to influence or observe a state variable in the system. Next, the similarity of the driver and sensor nodes is presented. The similarity of driver nodes x_4 and x_6 is 0.81. In this case, the reason why it is less than 1 is that although they control the same set of nodes, the numbers of derivations that influence

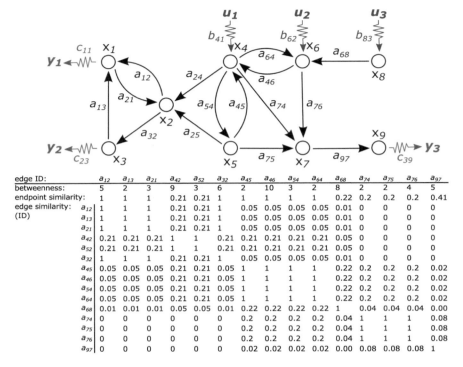

edge ID:		a_{12}	a_{13}	a_{21}	a_{42}	a_{52}	a_{32}	a_{45}	a_{46}	a_{54}	a_{64}	a_{68}	a_{74}	a_{75}	a_{76}	a_{97}
betweenness:		5	2	3	9	3	6	2	10	3	2	8	2	2	4	5
endpoint similarity:		1	1	1	0.21	0.21	1	1	1	1	1	0.22	0.2	0.2	0.2	0.41
edge similarity	a_{12}	1	1	1	0.21	0.21	1	0.05	0.05	0.05	0.05	0.01	0	0	0	0
(ID)	a_{13}	1	1	1	0.21	0.21	1	0.05	0.05	0.05	0.05	0.01	0	0	0	0
	a_{21}	1	1	1	0.21	0.21	1	0.05	0.05	0.05	0.05	0.01	0	0	0	0
	a_{42}	0.21	0.21	0.21	1	1	0.21	0.21	0.21	0.21	0.21	0.05	0	0	0	0
	a_{52}	0.21	0.21	0.21	1	1	0.21	0.21	0.21	0.21	0.21	0.05	0	0	0	0
	a_{32}	1	1	1	0.21	0.21	1	0.05	0.05	0.05	0.05	0.01	0	0	0	0
	a_{45}	0.05	0.05	0.05	0.21	0.21	0.05	1	1	1	1	0.22	0.2	0.2	0.2	0.02
	a_{46}	0.05	0.05	0.05	0.21	0.21	0.05	1	1	1	1	0.22	0.2	0.2	0.2	0.02
	a_{54}	0.05	0.05	0.05	0.21	0.21	0.05	1	1	1	1	0.22	0.2	0.2	0.2	0.02
	a_{64}	0.05	0.05	0.05	0.21	0.21	0.05	1	1	1	1	0.22	0.2	0.2	0.2	0.02
	a_{68}	0.01	0.01	0.01	0.05	0.05	0.01	0.22	0.22	0.22	0.22	1	0.04	0.04	0.04	0.00
	a_{74}	0	0	0	0	0	0	0.2	0.2	0.2	0.2	0.04	1	1	1	0.08
	a_{75}	0	0	0	0	0	0	0.2	0.2	0.2	0.2	0.04	1	1	1	0.08
	a_{76}	0	0	0	0	0	0	0.2	0.2	0.2	0.2	0.04	1	1	1	0.08
	a_{97}	0	0	0	0	0	0	0.02	0.02	0.02	0.02	0.00	0.08	0.08	0.08	1

Fig. 5.6 Calculated edge centrality measures for the given topology

them are different. In terms of sensor similarity, sensor nodes x_2 and x_3 observe the same set of nodes and they do this almost simultaneously, so their similarity is 0.91. R_C and R_O are the simple reachability matrices, respectively. They show which nodes can be controlled or observed by a given node. In R_C, the ith column shows which nodes can control node i. From the other viewpoint, elements in row i highlight those nodes which can be controlled by node i. In this example, node x_8 can influence every node, but it does not guarantee structural controllability. The R_O matrix can be interpreted analogously with regard to observability.

Finally, measures of edge centrality are seen in Fig. 5.6. The betweenness has the same meaning as in the case of nodes, that is, it yields the number of shortest paths that intercept the edge [7]. From this perspective, the most critical edge is the edge a_{46} with a value of 10. The endpoint similarity shows how similar the influenced and observed sets of the state variables with regard to the endpoints of edges are. This metric has a high value if the edge is part of a cycle or creates a bridge in the network. As no bridges are present in this network, only cycles can be recognised by this measure. The edge similarity shows how similar the roles of edges are, and it allows redundancies, to be located. In the topology presented, nodes x_1, x_2 and x_3, or nodes x_4, x_5, x_6 and x_7 also create parts of the network that possess redundancy.

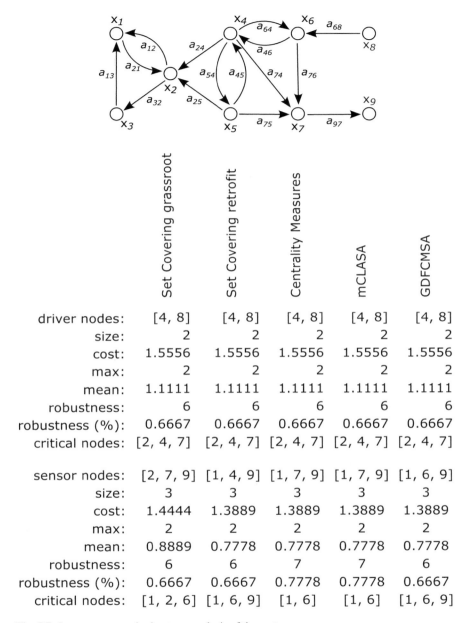

	Set Covering grassroot	Set Covering retrofit	Centrality Measures	mCLASA	GDFCMSA
driver nodes:	[4, 8]	[4, 8]	[4, 8]	[4, 8]	[4, 8]
size:	2	2	2	2	2
cost:	1.5556	1.5556	1.5556	1.5556	1.5556
max:	2	2	2	2	2
mean:	1.1111	1.1111	1.1111	1.1111	1.1111
robustness:	6	6	6	6	6
robustness (%):	0.6667	0.6667	0.6667	0.6667	0.6667
critical nodes:	[2, 4, 7]	[2, 4, 7]	[2, 4, 7]	[2, 4, 7]	[2, 4, 7]
sensor nodes:	[2, 7, 9]	[1, 4, 9]	[1, 7, 9]	[1, 7, 9]	[1, 6, 9]
size:	3	3	3	3	3
cost:	1.4444	1.3889	1.3889	1.3889	1.3889
max:	2	2	2	2	2
mean:	0.8889	0.7778	0.7778	0.7778	0.7778
robustness:	6	6	7	7	6
robustness (%):	0.6667	0.6667	0.7778	0.7778	0.6667
critical nodes:	[1, 2, 6]	[1, 6, 9]	[1, 6]	[1, 6]	[1, 6, 9]

Fig. 5.7 Improvement and robustness analysis of the system

For the demonstration of the last module, configurations provided by the first module are used again (Fig. 5.2). Results provided by this module can be seen in Fig. 5.7. In this case, five methods were applied to the system to extend the configuration as follows: the required relative degree was set at 2, while the alpha parameter of the cost function was set at 0.5 [9]. Results show that all the methods determine the same set of driver nodes for the system, that is, they are sufficient to influence state variables x_4 and x_8. The resultant cost is 1.5556, the relative degree is 2 which satisfies the requirements and the mean of the relative degrees is 1.1111. In this configuration, six different nodes can be identified which can be damaged separately and the system remains controllable. This is expressed by the value of robustness (66.6%). The most important nodes in terms of controllability are x_2, x_4 and x_7. In the case of observability, methods yield different solutions with the exception of the centrality measures-based and mCLASA algorithms which provide the best configuration in this case. Although the cost as well as the maximum and mean of the relative degree were identical in the case of retrofit set covering -based and GDFCMSA methods as well, the robustness analysis of these configurations exhibits a higher degree of vulnerability.

5.5   Discussion

In this chapter the Octave-and MATLAB-compatible NOCAD toolbox [1] was proposed to support the network-based controllability and observability analysis of dynamical systems. The toolbox offers two methods to design a structurally controllable and observable system based on the state-transition matrix. The designed system can be analysed by 49 qualitative measures both from structural and dynamical points of view. The toolbox serves five methods to improve the designed system by adding new inputs and outputs to it, thus, its relative degree can be decreased. Then the robustness of the individual designs can also be evaluated. The modular structure of the toolbox supports the facile improvement of the modules by adding new functions and the toolbox can be extended by new modules as well. Even though the modules are built on each other, most of their functions can also be used independently from each other.

Although our goal in this chapter is to draw the attention of researchers to the services provided by the NOCAD toolbox, it can be utilised in practice in various fields of sciences as well, for example, it enables social networks to be controlled in the economy, transaction networks to be analysed in finance or dynamical systems to be designed in engineering.

References

1. Abonyi, J.: abonyilab/nocad v2.0, May 2019
2. Borrett, S.R., Lau, M.K.: enaR: an R package for ecosystem network analysis. Methods Ecol. Evol. **5**(11), 1206–1213 (2014). https://github.com/SEELab/enaR. Accessed 8 Apr 2019
3. Bounova, G.: Octave networks toolbox (2015)

4. Chaturvedi, V.: Controllability of networks (2015). https://github.com/Vatshank/graph-control. Accessed 21 Mar 2019
5. Chu, Y., Wang, Z., Wang, R., Zhang, N., Li, J., Hu, Y., Teng, M., Wang, Y.: WDNfinder: a method for minimum driver node set detection and analysis in directed and weighted biological network. J. Bioinform. Comput. Biol. **15**(05), 1750021 (2017). https://github.com/dustincys/WDNfinder/blob/master/readme.md. Accessed 8 Apr 2019
6. Faradonbeh M.K.S., Tewari A., Michailidis G.: Optimality of fast-matching algorithms for random networks with applications to structural controllability. IEEE Trans. Control. Netw. Syst. **4**(4), 770–780 (2017)
7. Freeman, L.C.: A set of measures of centrality based on betweenness. Sociometry, 35–41 (1977)
8. Gori, F., Folino, G., Jetten, M.S.M., Marchiori, E.: MTR: taxonomic annotation of short metagenomic reads using clustering at multiple taxonomic ranks. Bioinformatics **27**(2), 196–203 (2010)
9. Leitold D., Vathy-Fogarassy Á., Abonyi J.: Network distance-based simulated annealing and fuzzy clustering for sensor placement ensuring observability and minimal relative degree. Sensors **18**(9), 3096 (2018)
10. Leitold D., Vathy-Fogarassy Á., Abonyi J.: Evaluation of the complexity, controllability and observability of heat exchanger networks based on structural analysis of network representations. Energies **12**(3):513 (2019)
11. Leitold, D., Vathy-Fogarassy, Á., Abonyi, J.: Network-based observability and controllability analysis of dynamical systems: the NOCAD toolbox [version 2; peer review: 2 approved]. F1000Research **8**, 646 (2019)
12. Liu Y.-Y., Barabási A.-L.: Control principles of complex systems. Rev. Mod. Phys. **88**(3), 035006 (2016)
13. Liu, Y.-Y., Slotine, J.-J., Barabási, A.-L.: Controllability of complex networks. Nature **473**(7346), 167 (2011)
14. Nepusz, T., Vicsek, T.: Controlling edge dynamics in complex networks. Nat. Phys. **8**(7), 568 (2012). https://github.com/ntamas/netctrl. Accessed 8 Apr 2019
15. Ruths J., Ruths D.: Control profiles of complex networks. Science **343**(6177), 1373–1376 (2014)
16. Shannon, P., Markiel, A., Ozier, O., Baliga, N.S., Wang, J.T., Ramage, D., Amin, N., Schwikowski, B., Ideker, T.: Cytoscape: a software environment for integrated models of biomolecular interaction networks. Genome Res. **13**(11), 2498–2504 (2003)
17. Taylor, A., Higham, D.J.: Contest: a controllable test matrix toolbox for matlab. ACM Trans. Math. Softw. (TOMS), **35**(4), 26 (2009). https://pureportal.strath.ac.uk/en/publications/contest-a-controllable-test-matrix-toolbox-for-matlab. Accessed 8 Apr 2019
18. Wu, L., Li, M., Wang, J., Wu, F-X.: Cytoctrlanalyser: a cytoscape app for biomolecular network controllability analysis. Bioinformatics **34**(8), 1428–1430 (2017). http://apps.cytoscape.org/apps/cytoctrlanalyser. Accessed 8 Apr 2019
19. Zinoviev, D.: Complex Network Analysis in Python: Recognize-Construct-Visualize-Analyze-Interpret. Pragmatic Bookshelf (2018)

Chapter 6
Summary

Abstract The application of network science in the field of dynamical systems has enabled a set of powerful tools that were introduced in this book. The typical structural motifs were determined for reaching more reliable results. The effect of these motifs on the number of driver and sensor nodes was also analysed. As the decreased number of inputs and outputs results in a higher relative degree, five methods were proposed to improve the operability performances by placing additional driver and sensor nodes to the system. A workflow for systematic analysis of dynamical systems and an open-source toolbox were also introduced to support the application of the presented methods. In this chapter, a summary of the book is given.

Keywords Dynamical systems · Controllability · Observability · Network science · Relative degree · Heat exchanger network

6.1 Conclusion

The network science-based analysis of dynamical systems has become a very popular research field over the past decade. The approach has also enabled the interpretation of dynamical properties in interconnected entities that are described by networks other than dynamical systems such as social networks, citation networks or food webs.

The importance of the appropriate representation of the connections between the elements in these systems has been increasing and come into focus. The most contestable point of the approach is that it is only based on a static and structural view of the system. As a result, usage of the proper topology of the analysed system is the most critical part of the methodology, since this is the only way to emphasise dynamics in static representations. As a solution, connection types were introduced according to the typical motifs of the state variables and nodes, that can represent the integrating behaviour of state variables, diffusion or information spreading in networks. The interpretation of these types in networks significantly reduced the number of driver and sensor nodes, and, not incidentally, increased the relative degree.

To grant fast and robust process monitoring and control, a centrality measures-based, two set covering-based as well as two clustering and simulated annealing-based methods were introduced, that can optimise the placement of the additional driver and sensor nodes. In other problems, the reduced relative degree can result in decreased spreading times of information in social networks, faster shipping in a transportation network or faster acting painkillers.

As heat exchanger networks (HENs) are thoroughly researched dynamical systems that contain typical structural patterns, a systematic workflow to analyse dynamical systems was given. The main characteristics of HEN design methods were determined and the effect of structural properties on dynamical behaviour was revealed with the aid of this workflow. The workflow can be utilised in the analysis of any kind of network. Even though some HEN-specific steps were included in our case study, these can be easily substituted or skipped. Although considerable research has utilised this methodology, a flexible software tool was missing to support such researches. To fill this gap, an Octave- and MATLAB-compatible NOCAD toolbox was proposed.

In the first part of the book, the typical structural motifs of dynamical systems were analysed. The interpretation of the revealed connection types in 35 benchmark networks was introduced, and its effect on the number of driver and sensor nodes was analysed. As a result, it was concluded that

- the majority (27 out of 35) of the networks studied in the literature do not exhibit dynamical behaviour that could be interpreted as a (linear) state-space model, even though diffusion in various networks or snapshots of temporal networks could ensure this,
- if the connection types were also taken into account, the number of necessary driver and sensor nodes would be drastically decreased (by more than 95%) compared to the analysis of oversimplified structural networks,
- network science-based analysis should follow the novel workflow proposed,
- the reduced number of inputs and outputs results in a system with a high relative degree that prevents fast and robust process monitoring and control.

To handle the increased relative degree as well as ensure fast and robust process monitoring, control, information spreading or transportation, five methods were proposed to place additional driver and sensor nodes in addition to the minimum ones defined by the path-finding method. Following performance analysis of the methods, it was concluded that

- centrality measures and set covering can also be utilised to determine the necessary additional sensors (and drivers),
- a cost function was defined that minimises the maximum and average of the relative degrees of the system to generate a balanced placement of sensors (and drivers),
- two simulated annealing-based methods were also proposed that can optimise the placement and solutions to massive problems generated by both approaches to illustrate its applicability in an industrial setting.

To demonstrate the applicability of the proposed methods, a systematic analysis of more than 50 heat integration problems was presented that were solved by in excess of 20 design methodologies. For the analysis, three network-based representations were presented. Although three network representations of HENs were utilised in the case studies, the elements of the proposed framework can easily be adapted to other fields.

The result of the analysis of more than 600 HENs revealed that

- the network science-based representations could be beneficial to reveal the complexity of HENs in terms of their operability,
- the proposed approach can be efficiently applied to the fast screening of HENs based on their structural properties,
- the popular maximum matching-based method used to determine the sensor and driver nodes in dynamical systems was unsuitable for heat exchanger networks due to its intrinsic dynamics,
- in more interconnected HENs, fewer actuators and sensors are needed to ensure structural controllability and observability, which results in a higher relative degree and increased difficulty of the operation,
- the degree-based structural measures can refer to the level of heat integration,
- the methodology was suitable for classifying the different methods that grant the HENs, as some techniques tended to determine more units in the network than others.

Last, but not least, an Octave- and MATLAB-compatible toolbox was introduced, that provides an open solution to support network science-based analysis referred to as the Network-based Observability and Controllability Analysis of Dynamical Systems (NOCAD) toolbox. The methods introduced in this book were implemented in the toolbox as well as further centrality and clustering measures. The toolbox provides

- the network-based controllability and observability analysis of any dynamical systems,
- two methods to design a structurally controllable and observable system based on the state-transition matrix,
- system analysis by 49 qualitative measures both from structural and dynamical points of view,
- five methods to improve the designed system by adding new inputs and outputs to it, thus, its relative degree can be decreased,
- robustness analysis of the individual designs,
- a modular structure that supports the facile improvement of the modules by adding new functions and the toolbox can be extended by new modules as well.

Thus, even though in our book the methods were introduced through particular types of dynamical systems, the toolbox facilitates the analysis of dynamical and networked systems, e.g. social networks with dynamics originating from information, communication or knowledge diffusion; transportation networks; or chemical and biological systems described by several connected state variables.

Appendix A
Effect of Connection Types

This chapter contains additional information for Chap. 2.

A.1   The Path-Finding Method

In special cases the maximum matching algorithm provides such result, in which no unmatched nodes are identified, so the system is not controllable and/or observable. In the literature, the number of driver nodes is denoted by N_D and $N_D = max\{N - |M^*|, 1\}$, where $N - |M^*|$ is the number of unmatched nodes. When the number of the unmatched nodes is equal to 0, then theoretically, the whole system can be controlled by a driver node, independently from its location. This phenomenon can be true for a part of the system as well.

Liu et al. proposed an approach to solve this problem. To achieve the controllability, they share the signal of an input on each uncontrolled part of the system, i.e. they influence more state variables with only one input [11]. Since these uncontrolled parts have self-regulating property, it is enough to influence an arbitrary input with a signal, i.e. with an input we can control an unmatched node and uncontrolled subsystems as well. According to Liu et al. this problem is caused by uncontrolled strongly connected components (SCC). It is true, since each node is matched by a previous one, so we can draw a cycle by the matched nodes and this creates an SCC. Since there are SCCs where unmatched nodes are identified, we examine more deeply what phenomenon results in the uncontrolled parts. This was motivated by the lack of appropriate cause, and by the fact, that sharing an input signal is not feasible in every dynamical system.

As a result, we found that, these uncontrolled parts are resulted by Hamiltonian cycles. To eliminate the sharing of an input signal, we have developed a novel method, that cuts each Hamiltonian cycle and creates a Hamiltonian path so that, if it is possible, then the path connects to an unmatched node. With the elimination of an edge

© The Author(s), under exclusive license to Springer Nature Switzerland AG 2020
D. Leitold et al., *Network-Based Analysis of Dynamical Systems*,
SpringerBriefs in Computer Science,
https://doi.org/10.1007/978-3-030-36472-4

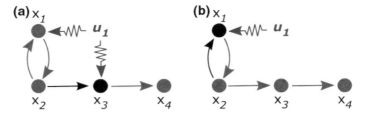

Fig. A.1 An illustrative example with input configurations (u) generated by the **a** signal sharing and the **b** path-finding methods. Red lines show the matched edges. The unmatched node in **a** is x_3, and in **b** is x_1. In the case of the signal sharing method, node x_3 is controlled by u_1 and uncontrolled SCC $\{x_1, x_2\}$ should be also controlled by this input. In contrast, path-finding method modifies the disjoint edge set by the exchange the edge (x_2, x_1) to the edge (x_2, x_3). By changing the result of the maximum matching algorithm, the path-finding method can control the system without the sharing of the input signal

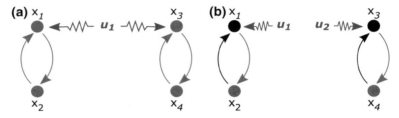

Fig. A.2 The input configurations (u) for an unconnected system generated by a signal sharing and by **b** path-finding methods. Red lines show the matched edges. Since the path-finding method changes the maximum matching algorithm to produce unmatched nodes, it can be that the new matching is not a maximum matching. It is clearly visible that the unconnected components increase the number of inputs in case of the path-finding method. In contrast, the signal sharing method shares the signal of the only input, u_1

from the cycle, and taking another edge into the disjoint edge set we have exchanged the original maximum matching algorithm. Unfortunately, if there is not such an unmatched node, then only the cycle is modified, i.e. we remove an edge, so the resulted matching is not a maximum matching. Since the method finds Hamiltonian paths, we called this new method as *path-finding* method, and the original one, as it shares signal, *signal sharing* method. Results of the presented two methods are seen in Fig. A.1.

Although path-finding method simplifies the controlling process, it has disadvantages as well. Since the result of the maximum matching algorithm is changed, the number of driver nodes can increase, i.e. it does not provide a solution with minimal number of inputs to control the system. But, in cases, when we want to control the system without sharing of input signals, the proposed path-finding method provides inputs with minimal number. Thus, unconnected components in the system result in more inputs in the case of path-finding method, than in the case of signal sharing method (Fig. A.2).

A.2 The Studied Networks

The networks used as a set of benchmark examples are seen in Table A.1. *Network Set I* contains networks, which are used in controllability examinations and their dynamical behavior is interpretable. In contrast, topologies in *Network Set II* describe processes in which dynamical behaviors are not interpretable. *Network Set III* contains topologies which originated from state-transition matrices of real dynamical systems.

A.3 Calculated Measures of the Benchmark Networks

In Tables A.2, A.3, A.4 and A.5, the generated measures can be seen for the presented networks with influence, self-affecting influence, interaction and self-affecting interaction connection types, respectively.

Table A.1 Topologies used in the study. Network Set I contains 8 example networks, Network Set II contains 27 example networks and Network Set III contains 18 example networks. The table shows the names, and the descriptions of the networks, and N yields the number of nodes and $|E|$ stands for number of edges

Type	Name	N	$|E|$	Description (Citation)
Network Set I	C. elegans	297	2,345	C. elegans neural network [20]
	ecoli	418	519	Transcriptional regulation network of E. coli [18]
	mac71	71	746	Macaque monkey visual cortex [21]
	mac95	94	2,390	Macaque monkey visual cortex [5]
	s208	122	189	Power grid network [13]
	s420	252	399	Power grid network [13]
	s838	512	819	Power grid network [13]
	yeast	688	1,079	Protein-protein interaction network in yeast [14]
Network Set II	amazon0302	262,111	1,234,877	Amazon co-purchasing network, March 2 2003 [6]
	amazon0312	400,727	3,200,440	Amazon co-purchasing network, March 12 2003 [6]
	amazon0505	410,236	3,356,824	Amazon co-purchasing network, May 5 2003 [6]
	amazon0601	403,394	3,387,388	Amazon co-purchasing network, June 1 2003 [8]
	berkstan	26,475	106,762	AS graph from RouteViews BGP table snapshots [8]
	caida	685,230	7,600,595	Web graph of Berkeley and Stanford [10]
	dolphins	62	159	Social network of bottlenose dolphins. [12]
	epinion	75,879	508,837	Who-trusts-whom network of Epinions.com [17]
	freeman2	34	830	Intra organization network [4]
	gnutella04	10,876	39,994	Gnutella peer to peer network, August 4 200. [9]
	gnutella05	8,846	31,839	Gnutella peer to peer network, August 5 2002 [9]
	gnutella06	8,717	31,525	Gnutella peer to peer network, August 6 2002 [9]
	gnutella08	6,301	20,777	Gnutella peer to peer network, August 8 2002 [9]
	gnutella09	8,114	26,013	Gnutella peer to peer network, August 9 2002 [9]

(continued)

Table A.1 (continued)

	gnutella24	26,518	65,369	Gnutella peer to peer network, August 24 2002 [17]
	gnutella25	22,687	54,705	Gnutella peer to peer network, August 25 2002 [17]
	gnutella30	36,682	88,328	Gnutella peer to peer network, August 30 2002 [17]
	gnutella31	62,586	147,892	Gnutella peer to peer network, August 31 2002 [17]
	google	875,713	5,105,039	Web graph from Google [10]
	grass	88	137	Food web of Grassland [3]
	notre	325,729	1,497,134	Web graph of Notre Dame [1]
	pokec	1,632,803	30,622,564	Pokec online social network [19]
	slash08	77,360	905,468	Slashdot Zoo social network, November 6 2008 [10]
	slash09	82,168	948,464	Slashdot Zoo social network, February 21 2009 [10]
	stanford	281,903	2,312,497	Web graph of Stanford.edu [10]
	vote	7,115	103,689	Wikipedia who-votes-on-whom network [7]
	ythan	135	601	Food web of Ythan Estuary [3]
Network Set III	beam	348	60,726	Clamped beam model [2]
	build	48	1,176	Motion problem in a building [2]
	CDplayer	120	240	Classical CD player model [2]
	eady	598	357,406	Model of an atmospheric storm track [2]
	fom	1,006	1,012	Described in cited article [15]
	heatCont	200	598	Heat equation in a thin rod [2]
	heatDisc	200	598	Discretization of the previous equation [2]
	iss	270	405	Component 1r of the International Space Station [2]
	MNA_1	578	1,694	Modified Nodal Analysis model [2]
	MNA_2	9,223	27,003	Modified Nodal Analysis model [2]
	MNA_3	4,863	13,921	Modified Nodal Analysis model [2]
	MNA_4	980	2,872	Modified Nodal Analysis model [2]
	MNA_5	10,913	54,159	Modified Nodal Analysis model [2]
	orrSom	100	10,000	Orr-Sommerfeld operator for Couette flow [2]
	pde	84	382	Partial differential equation [2]
	PEEC	480	1,346	Partial element equivalent circuit model [2]
	random	200	2,132	Random example [2]
	tline	256	256	Example of a transmission line model [2]

Table A.2 Results of examinations of networks with influence dynamic. In header, N yields the number of nodes, $|E|$ yields the number of edges, D yields the density, $<k>$ yields the average degree, N_D yields the number of driver nodes, N_S yields the number of sensor nodes, $\%_D$ yields the proportion of driver and all nodes, $\%_S$ yields the proportion of sensor and all nodes, $\%_{loop}$ yields the percentage of loops and $\%_{sym}$ yields the percentage of symmetries

Network	N	$|E|$	D	$<k>$	N_D	N_S	$\%_D$	$\%_S$	$\%_{loop}$	$\%_{sym}$
C.elegans	297	2345	0.0266	15.7912	50	49	16.8350	16.4983	0.0000	16.8017
ecoli	418	519	0.0030	2.4833	314	314	75.1196	75.1196	0.0000	0.0000
mac71	71	746	0.1480	21.0141	1	1	1.4085	1.4085	0.0000	82.5737
mac95	94	2390	0.2705	50.8511	9	9	9.5745	9.5745	0.0000	73.2218
s208	122	189	0.0127	3.0984	29	29	23.7705	23.7705	0.0000	0.0000
s420	252	399	0.0063	3.1667	59	59	23.4127	23.4127	0.0000	0.0000
s838	512	819	0.0031	3.1992	119	119	23.2422	23.2422	0.0000	0.0000
yeast	688	1079	0.0023	3.1366	565	565	82.1221	82.1221	0.0000	0.1854
amazon0302	262111	1234877	0.0000	9.4225	8458	9238	3.2269	3.5245	0.0000	54.2702
amazon0312	400727	3200440	0.0000	15.9732	14103	14382	3.5194	3.5890	0.0000	53.1534
amazon0505	410236	3356824	0.0000	16.3653	14840	15062	3.6174	3.6715	0.0000	54.6580
amazon0601	403394	3387388	0.0000	16.7944	8294	8398	2.0561	2.0818	0.0000	55.7350
berkstan	685230	7600595	0.0000	22.1841	261570	261687	38.1726	38.1897	0.0000	25.0276
caida	26475	106762	0.0002	8.0651	19112	19112	72.1889	72.1889	0.0000	100.0000
dolphins	62	159	0.0414	5.1290	19	19	30.6452	30.6452	0.0000	0.0000
epinion	75879	508837	0.0001	13.4118	41889	41792	55.2050	55.0772	0.0000	40.5226
freeman2	34	830	0.7180	48.8235	1	1	2.9412	2.9412	0.0000	85.7831
gnutella04	10876	39994	0.0003	7.3545	6004	6004	55.2041	55.2041	0.0000	0.0000
gnutella05	8846	31839	0.0004	7.1985	5111	5111	57.7775	57.7775	0.0000	0.0000
gnutella06	8717	31525	0.0004	7.2330	5033	5033	57.7378	57.7378	0.0000	0.0000
gnutella08	6301	20777	0.0005	6.5948	4106	4106	65.1643	65.1643	0.0000	0.0000
gnutella09	8114	26013	0.0004	6.4119	5355	5355	65.9970	65.9970	0.0000	0.0000
gnutella24	26518	65369	0.0001	4.9302	18965	18965	71.5175	71.5175	0.0000	0.0000
gnutella25	22687	54705	0.0001	4.8226	16478	16478	72.6319	72.6319	0.0000	0.0000
gnutella30	36682	88328	0.0001	4.8159	26965	26965	73.5102	73.5102	0.0000	0.0000

(continued)

Table A.2 (continued)

Network	N	$|E|$	D	$<k>$	N_D	N_S	$\%_D$	$\%_S$	$\%_{loop}$	$\%_{sym}$
gnutella31	62586	147892	0.0000	4.7260	46227	46227	73.8616	73.8616	0.0000	0.0000
google	875713	5105039	0.0000	11.6592	423003	423964	48.3038	48.4136	0.0000	30.6751
grass	88	137	0.0177	3.1136	46	46	52.2727	52.2727	0.0000	0.0000
notre	325729	1497134	0.0000	9.0239	220552	221769	67.7103	68.0839	8.4288	51.6536
pokec	1632803	30622564	0.0000	37.5092	252075	252195	15.4382	15.4455	0.0000	54.3429
slash08	77360	905468	0.0002	21.4106	49	6698	0.0633	8.6582	99.9315	86.6935
slash09	82168	948464	0.0001	21.1800	3737	10495	4.5480	12.7726	95.2962	84.1065
stanford	281903	2312497	0.0000	16.4063	90168	90211	31.9855	32.0007	0.0000	27.6637
vote	7115	103689	0.0020	29.1466	4736	4736	66.5636	66.5636	0.0000	5.6457
ythan	135	601	0.0330	8.8444	69	69	51.1111	51.1111	2.9630	0.3350
beam	348	60726	0.5014	348.0000	1	1	0.2874	0.2874	50.0000	50.2874
build	48	1176	0.5104	48.0000	1	1	2.0833	2.0833	50.0000	52.0833
CDplayer	120	240	0.0167	2.0000	60	60	50.0000	50.0000	100.0000	100.0000
eady	598	357406	0.9994	1193.3378	1	1	0.1672	0.1672	100.0000	99.9462
fom	1006	1012	0.0010	0.0119	1003	1003	99.7018	99.7018	100.0000	100.0000
heatCont	200	598	0.0150	3.9800	1	1	0.5000	0.5000	100.0000	100.0000
heatDisc	200	598	0.0150	3.9800	1	1	0.5000	0.5000	100.0000	100.0000
iss	270	405	0.0056	2.0000	135	135	50.0000	50.0000	50.0000	100.0000
MNA_1	578	1694	0.0051	4.6713	3	3	0.5190	0.5190	59.5156	100.0000
MNA_2	9223	27003	0.0003	4.6358	4	4	0.0434	0.0434	60.9888	100.0000
MNA_3	4863	13921	0.0006	4.5116	9	9	0.1851	0.1851	60.6827	100.0000
MNA_4	980	2872	0.0030	4.6449	2	2	0.2041	0.2041	60.8163	100.0000
MNA_5	10913	54159	0.0005	7.9417	10	10	0.0916	0.0916	99.1936	100.0000
orrSom	100	10000	1.0000	198.0000	1	1	1.0000	1.0000	100.0000	100.0000
pde	84	382	0.0541	7.0952	1	1	1.1905	1.1905	100.0000	100.0000
PEEC	480	1346	0.0058	4.3417	1	1	0.2083	0.2083	63.3333	100.0000
random	200	2132	0.0533	19.3200	1	1	0.5000	0.5000	100.0000	4.2443
tline	256	256	0.0039	0.0000	256	256	100.0000	100.0000	100.0000	0.0000

Table A.3 Results of examinations of networks with self-affecting influence dynamic. In header, N yields the number of nodes, $|E|$ yields the number of edges, D yields the density, $< k >$ yields the average degree, N_D yields the number of driver nodes, N_S yields the number of sensor nodes, $\%_D$ yields the proportion of driver and all nodes, $\%_S$ yields the proportion of sensor and all nodes, $\%_{loop}$ yields the percentage of loops and $\%_{sym}$ stands for the percentage of symmetries

Network	N	$|E|$	D	$< k >$	N_D	N_S	$\%_D$	$\%_S$	$\%_{loop}$	$\%_{sym}$
C.elegans	297	2642	0.0300	15.7912	28	3	9.4276	1.0101	100.0000	16.8017
ecoli	418	937	0.0054	2.4833	312	76	74.6411	18.1818	100.0000	0.0000
mac71	71	817	0.1621	21.0141	1	1	1.4085	1.4085	100.0000	82.5737
mac95	94	2484	0.2811	50.8519	9	1	9.5745	1.0638	100.0000	73.2218
s208	122	311	0.0209	3.0984	10	1	8.1967	0.8197	100.0000	0.0000
s420	252	651	0.0103	3.1667	18	1	7.1429	0.3968	100.0000	0.0000
s838	512	1331	0.0051	3.1992	34	1	6.6406	0.1953	100.0000	0.0000
yeast	688	1767	0.0037	3.1366	96	557	13.9535	80.9593	100.0000	0.1854
amazon0302	262111	1496988	0.0000	9.4225	1	5668	0.0004	2.1624	100.0000	54.2702
amazon0312	400727	3601167	0.0000	15.9732	1	12705	0.0002	3.1705	100.0000	53.1534
amazon0505	410236	3767060	0.0000	16.3653	2	13712	0.0005	3.3425	100.0000	54.6580
amazon0601	403394	3790782	0.0000	16.7944	155	1260	0.0384	0.3123	100.0000	55.7350
berkstan	685230	8285825	0.0000	22.1841	70171	8135	10.2405	1.1872	100.0000	25.0276
caida	26475	133237	0.0002	8.0651	1	1	0.0038	0.0038	100.0000	100.0000
dolphins	62	221	0.0575	5.1290	15	12	24.1935	19.3548	100.0000	0.0000
epinion	75879	584716	0.0001	13.4118	24377	15868	32.1261	20.9122	100.0000	40.5226
freeman2	34	864	0.7474	48.8235	1	1	2.9412	2.9412	100.0000	85.7831
gnutella04	10876	50870	0.0004	7.3545	20	5941	0.1839	54.6249	100.0000	0.0000
gnutella05	8846	40685	0.0005	7.1985	118	4996	1.3339	56.4775	100.0000	0.0000
gnutella06	8717	40242	0.0005	7.2330	79	4978	0.9063	57.1068	100.0000	0.0000
gnutella08	6301	27078	0.0007	6.5948	80	3836	1.2696	60.8792	100.0000	0.0000
gnutella09	8114	34127	0.0005	6.4119	76	5059	0.9367	62.3490	100.0000	0.0000
gnutella24	26518	91887	0.0001	4.9302	331	18948	1.2482	71.4534	100.0000	0.0000
gnutella25	22687	77392	0.0002	4.8226	335	16466	1.4766	72.5790	100.0000	0.0000
gnutella30	36682	125010	0.0001	4.8159	229	26960	0.6243	73.4965	100.0000	0.0000
gnutella31	62586	210478	0.0001	4.7260	303	46199	0.4841	73.8168	100.0000	0.0000
google	875713	5980752	0.0000	11.6592	162465	141104	18.5523	16.1130	100.0000	30.6751
grass	88	225	0.0291	3.1136	1	35	1.1364	39.7727	100.0000	0.0000
notre	325729	1795408	0.0000	9.0239	1	189150	0.0003	58.0697	100.0000	51.6536
pokec	1632803	32255367	0.0000	37.5092	114165	201024	6.9920	12.3116	100.0000	54.3429
slash08	77360	905521	0.0002	21.4106	1	6698	0.0013	8.6582	100.0000	86.6935
slash09	82168	952329	0.0001	21.1800	1	10495	0.0012	12.7726	100.0000	84.1065
stanford	281903	2594400	0.0000	16.4063	21410	2403	7.5948	0.8524	100.0000	27.6637
vote	7115	110804	0.0022	29.1466	4734	1005	66.5355	14.1251	100.0000	5.6457
ythan	135	732	0.0402	8.8444	1	52	0.7407	38.5185	100.0000	0.3350
beam	348	60900	0.5029	348.0000	1	1	0.2874	0.2874	100.0000	50.2874
build	48	1200	0.5208	48.0000	1	1	2.0833	2.0833	100.0000	52.0833
CDplayer	120	240	0.0167	2.0000	60	60	50.0000	50.0000	100.0000	100.0000
eady	598	357406	0.9994	1193.3378	1	1	0.1672	0.1672	100.0000	99.9462
fom	1006	1012	0.0010	0.0119	1003	1003	99.7018	99.7018	100.0000	100.0000
heatCont	200	598	0.0150	3.9800	1	1	0.5000	0.5000	100.0000	100.0000
heatDisc	200	598	0.0150	3.9800	1	1	0.5000	0.5000	100.0000	100.0000

(continued)

Table A.3　(continued)

Network	N	|E|	D	< k >	N_D	N_S	%$_D$	%$_S$	%$_{loop}$	%$_{sym}$
iss	270	540	0.0074	2.0000	135	135	50.0000	50.0000	100.0000	100.0000
MNA_1	578	1928	0.0058	4.6713	3	3	0.5190	0.5190	100.0000	100.0000
MNA_2	9223	30601	0.0004	4.6358	4	4	0.0434	0.0434	100.0000	100.0000
MNA_3	4863	15833	0.0007	4.5116	9	9	0.1851	0.1851	100.0000	100.0000
MNA_4	980	3256	0.0034	4.6449	2	2	0.2041	0.2041	100.0000	100.0000
MNA_5	10913	54247	0.0005	7.9417	10	10	0.0916	0.0916	100.0000	100.0000
orrSom	100	10000	1.0000	198.0000	1	1	1.0000	1.0000	100.0000	100.0000
pde	84	382	0.0541	7.0952	1	1	1.1905	1.1905	100.0000	100.0000
PEEC	480	1522	0.0066	4.3417	1	1	0.2083	0.2083	100.0000	100.0000
random	200	2132	0.0533	19.3200	1	1	0.5000	0.5000	100.0000	4.2443
tline	256	256	0.0039	0.0000	256	256	100.0000	100.0000	100.0000	0.0000

Table A.4　**Results of examinations of networks with interaction dynamic.** In header, N yields the number of nodes, $|E|$ yields the number of edges, D yields the density, $< k >$ yields the average degree, N_D yields the number of driver nodes, N_S yields the number of sensor nodes, %$_D$ yields the proportion of driver and all nodes, %$_S$ yields the proportion of sensor and all nodes, %$_{loop}$ yields the percentage of loops and %$_{sym}$ stands for the percentage of symmetries

Network	N	|E|	D	< k >	N_D	N_S	%$_D$	%$_S$	%$_{loop}$	%$_{sym}$
C.elegans	297	4296	0.0487	28.9293	14	14	4.7138	4.7138	0.0000	100.0000
ecoli	418	1038	0.0059	4.9665	233	233	55.7416	55.7416	0.0000	100.0000
mac71	71	876	0.1738	24.6761	1	1	1.4085	1.4085	0.0000	100.0000
mac95	94	3030	0.3429	64.4681	3	3	3.1915	3.1915	0.0000	100.0000
s208	122	378	0.0254	6.1967	1	1	0.8197	0.8197	0.0000	100.0000
s420	252	798	0.0126	6.3333	3	3	1.1905	1.1905	0.0000	100.0000
s838	512	1638	0.0062	6.3984	11	11	2.1484	2.1484	0.0000	100.0000
yeast	688	2156	0.0046	6.2674	450	450	65.4070	65.4070	0.0000	100.0000
amazon0302	262111	1799584	0.0000	13.7315	442	442	0.1686	0.1686	0.0000	100.0000
amazon0312	400727	4699738	0.0000	23.4561	2533	2533	0.6321	0.6321	0.0000	100.0000
amazon0505	410236	4878874	0.0000	23.7857	2658	2658	0.6479	0.6479	0.0000	100.0000
amazon0601	403394	4886816	0.0000	24.2285	823	823	0.2040	0.2040	0.0000	100.0000
berkstan	685230	13298940	0.0000	38.8160	191369	191369	27.9277	27.9277	0.0000	100.0000
caida	26475	106762	0.0002	8.0651	19112	19112	72.1889	72.1889	0.0000	100.0000
dolphins	62	318	0.0827	10.2581	2	2	3.2258	3.2258	0.0000	100.0000
epinion	75879	811480	0.0001	21.3888	31720	31720	41.8034	41.8034	0.0000	100.0000
freeman2	34	948	0.8201	55.7647	1	1	2.9412	2.9412	0.0000	100.0000
gnutella04	10876	79988	0.0007	14.7091	2180	2180	20.0441	20.0441	0.0000	100.0000
gnutella05	8846	63678	0.0008	14.3970	1992	1992	22.5187	22.5187	0.0000	100.0000
gnutella06	8717	63050	0.0008	14.4660	1907	1907	21.8768	21.8768	0.0000	100.0000
gnutella08	6301	41554	0.0010	13.1897	2194	2194	34.8199	34.8199	0.0000	100.0000
gnutella09	8114	52026	0.0008	12.8238	2971	2971	36.6157	36.6157	0.0000	100.0000
gnutella24	26518	130738	0.0002	9.8603	12111	12111	45.6709	45.6709	0.0000	100.0000
gnutella25	22687	109410	0.0002	9.6452	10665	10665	47.0093	47.0093	0.0000	100.0000
gnutella30	36682	176656	0.0001	9.6318	18153	18153	49.4875	49.4875	0.0000	100.0000
gnutella31	62586	295784	0.0001	9.4521	31209	31209	49.8658	49.8658	0.0000	100.0000
google	875713	8644102	0.0000	19.7419	268028	268028	30.6068	30.6068	0.0000	100.0000

(continued)

Table A.4 (continued)

Network	N	$\lvert E\rvert$	D	$<k>$	N_D	N_S	$\%_D$	$\%_S$	$\%_{loop}$	$\%_{sym}$
grass	88	274	0.0354	6.2273	22	22	25.0000	25.0000	0.0000	100.0000
notre	325729	2207671	0.0000	13.3867	188276	188276	57.8014	57.8014	8.4288	100.0000
pokec	1632803	44603928	0.0000	54.6348	70671	70671	4.3282	4.3282	0.0000	100.0000
slash08	77360	1015667	0.0002	24.2596	5	5	0.0065	0.0065	99.9315	100.0000
slash09	82168	1086763	0.0002	24.5463	71	71	0.0864	0.0864	95.2962	100.0000
stanford	281903	3985272	0.0001	28.2741	64488	64488	22.8760	22.8760	0.0000	100.0000
vote	7115	201524	0.0040	56.6476	2637	2637	37.0625	37.0625	0.0000	100.0000
ythan	135	1196	0.0656	17.6593	23	23	17.0370	17.0370	2.9630	100.0000
beam	348	90828	0.7500	521.0000	1	1	0.2874	0.2874	50.0000	100.0000
build	48	1728	0.7500	71.0000	1	1	2.0833	2.0833	50.0000	100.0000
CDplayer	120	240	0.0167	2.0000	60	60	50.0000	50.0000	100.0000	100.0000
eady	598	357598	1.0000	1193.9799	1	1	0.1672	0.1672	100.0000	100.0000
fom	1006	1012	0.0010	0.0119	1003	1003	99.7018	99.7018	100.0000	100.0000
heatCont	200	598	0.0150	3.9800	1	1	0.5000	0.5000	100.0000	100.0000
heatDisc	200	598	0.0150	3.9800	1	1	0.5000	0.5000	100.0000	100.0000
iss	270	405	0.0056	2.0000	135	135	50.0000	50.0000	50.0000	100.0000
MNA_1	578	1694	0.0051	4.6713	3	3	0.5190	0.5190	59.5156	100.0000
MNA_2	9223	27003	0.0003	4.6358	4	4	0.0434	0.0434	60.9888	100.0000
MNA_3	4863	13921	0.0006	4.5116	9	9	0.1851	0.1851	60.6827	100.0000
MNA_4	980	2872	0.0030	4.6449	2	2	0.2041	0.2041	60.8163	100.0000
MNA_5	10913	54159	0.0005	7.9417	10	10	0.0916	0.0916	99.1936	100.0000
orrSom	100	10000	1.0000	198.0000	1	1	1.0000	1.0000	100.0000	100.0000
pde	84	382	0.0541	7.0952	1	1	1.1905	1.1905	100.0000	100.0000
PEEC	480	1346	0.0058	4.3417	1	1	0.2083	0.2083	63.3333	100.0000
random	200	3982	0.0996	37.8200	1	1	0.5000	0.5000	100.0000	100.0000
tline	256	256	0.0039	0.0000	256	256	100.0000	100.0000	100.0000	0.0000

Table A.5 **Results of examinations of networks with self-affecting interaction dynamic.** In header, N yields the number of nodes, $\lvert E\rvert$ yields the number of edges, D yields the density, $<k>$ yields the average degree, N_D yields the number of driver nodes, N_S yields the number of sensor nodes, $\%_D$ yields the proportion of driver and all nodes, $\%_S$ yields the proportion of sensor and all nodes, $\%_{loop}$ yields the percentage of loops and $\%_{sym}$ yields the percentage of symmetries

Network	N	$\lvert E\rvert$	D	$<k>$	N_D	N_S	$\%_D$	$\%_S$	$\%_{loop}$	$\%_{sym}$
C.elegans	297	4593	0.0521	28.9293	1	1	0.3367	0.3367	100.0000	100.0000
ecoli	418	1456	0.0083	4.9665	29	29	6.9378	6.9378	100.0000	100.0000
mac71	71	947	0.1879	24.6761	1	1	1.4085	1.4085	100.0000	100.0000
mac95	94	3124	0.3536	64.4681	1	1	1.0638	1.0638	100.0000	100.0000
s208	122	500	0.0336	6.1967	1	1	0.8197	0.8197	100.0000	100.0000
s420	252	1050	0.0165	6.3333	1	1	0.3968	0.3968	100.0000	100.0000
s838	512	2150	0.0082	6.3984	1	1	0.1953	0.1953	100.0000	100.0000
yeast	688	2844	0.0060	6.2674	11	11	1.5988	1.5988	100.0000	100.0000
amazon0302	262111	2061695	0.0000	13.7315	1	1	0.0004	0.0004	100.0000	100.0000
amazon0312	400727	5100465	0.0000	23.4561	1	1	0.0002	0.0002	100.0000	100.0000
amazon0505	410236	5289110	0.0000	23.7857	1	1	0.0002	0.0002	100.0000	100.0000
amazon0601	403394	5290210	0.0000	24.2285	7	7	0.0017	0.0017	100.0000	100.0000

(continued)

Table A.5 (continued)

Network	N	$|E|$	D	$<k>$	N_D	N_S	$\%_D$	$\%_S$	$\%_{loop}$	$\%_{sym}$
berkstan	685230	13984170	0.0000	38.8160	676	676	0.0987	0.0987	100.0000	100.0000
caida	26475	133237	0.0002	8.0651	1	1	0.0038	0.0038	100.0000	100.0000
dolphins	62	380	0.0989	10.2581	1	1	1.6129	1.6129	100.0000	100.0000
epinion	75879	887359	0.0002	21.3888	2	2	0.0026	0.0026	100.0000	100.0000
freeman2	34	982	0.8495	55.7647	1	1	2.9412	2.9412	100.0000	100.0000
gnutella04	10876	90864	0.0008	14.7091	1	1	0.0092	0.0092	100.0000	100.0000
gnutella05	8846	72524	0.0009	14.3970	3	3	0.0339	0.0339	100.0000	100.0000
gnutella06	8717	71767	0.0009	14.4660	1	1	0.0115	0.0115	100.0000	100.0000
gnutella08	6301	47855	0.0012	13.1897	2	2	0.0317	0.0317	100.0000	100.0000
gnutella09	8114	60140	0.0009	12.8238	6	6	0.0739	0.0739	100.0000	100.0000
gnutella24	26518	157256	0.0002	9.8603	11	11	0.0415	0.0415	100.0000	100.0000
gnutella25	22687	132097	0.0003	9.6452	13	13	0.0573	0.0573	100.0000	100.0000
gnutella30	36682	213338	0.0002	9.6318	12	12	0.0327	0.0327	100.0000	100.0000
gnutella31	62586	358370	0.0001	9.4521	12	12	0.0192	0.0192	100.0000	100.0000
google	875713	9519815	0.0000	19.7419	2746	2746	0.3136	0.3136	100.0000	100.0000
grass	88	362	0.0467	6.2273	1	1	1.1364	1.1364	100.0000	100.0000
notre	325729	2505945	0.0000	13.3867	1	1	0.0003	0.0003	100.0000	100.0000
pokec	1632803	46236731	0.0000	54.6348	1	1	0.0001	0.0001	100.0000	100.0000
slash08	77360	1015720	0.0002	24.2596	1	1	0.0013	0.0013	100.0000	100.0000
slash09	82168	1090628	0.0002	24.5463	1	1	0.0012	0.0012	100.0000	100.0000
stanford	281903	4267175	0.0001	28.2741	365	365	0.1295	0.1295	100.0000	100.0000
vote	7115	208639	0.0041	56.6476	24	24	0.3373	0.3373	100.0000	100.0000
ythan	135	1327	0.0728	17.6593	1	1	0.7407	0.7407	100.0000	100.0000
beam	348	91002	0.7514	521.0000	1	1	0.2874	0.2874	100.0000	100.0000
build	48	1752	0.7604	71.0000	1	1	2.0833	2.0833	100.0000	100.0000
CDplayer	120	240	0.0167	2.0000	60	60	50.0000	50.0000	100.0000	100.0000
eady	598	357598	1.0000	1193.9799	1	1	0.1672	0.1672	100.0000	100.0000
fom	1006	1012	0.0010	0.0119	1003	1003	99.7018	99.7018	100.0000	100.0000
heatCont	200	598	0.0150	3.9800	1	1	0.5000	0.5000	100.0000	100.0000
heatDisc	200	598	0.0150	3.9800	1	1	0.5000	0.5000	100.0000	100.0000
iss	270	540	0.0074	2.0000	135	135	50.0000	50.0000	100.0000	100.0000
MNA_1	578	1928	0.0058	4.6713	3	3	0.5190	0.5190	100.0000	100.0000
MNA_2	9223	30601	0.0004	4.6358	4	4	0.0434	0.0434	100.0000	100.0000
MNA_3	4863	15833	0.0007	4.5116	9	9	0.1851	0.1851	100.0000	100.0000
MNA_4	980	3256	0.0034	4.6449	2	2	0.2041	0.2041	100.0000	100.0000
MNA_5	10913	54247	0.0005	7.9417	10	10	0.0916	0.0916	100.0000	100.0000
orrSom	100	10000	1.0000	198.0000	1	1	1.0000	1.0000	100.0000	100.0000
pde	84	382	0.0541	7.0952	1	1	1.1905	1.1905	100.0000	100.0000
PEEC	480	1522	0.0066	4.3417	1	1	0.2083	0.2083	100.0000	100.0000
random	200	3982	0.0996	37.8200	1	1	0.5000	0.5000	100.0000	100.0000
tline	256	256	0.0039	0.0000	256	256	100.0000	100.0000	100.0000	0.0000

References

1. Albert, Réka, Jeong, Hawoong, Barabási, Albert-László: Internet: diameter of the world-wide web. Nature **401**(6749), 130–131 (1999)
2. Chahlaoui, Y., Van Dooren, P.: A collection of benchmark examples for model reduction of linear time invariant dynamical systems (2002)
3. Dunne, J.A., Williams, R.J., Martinez, N.D.: Food-web structure and network theory: the role of connectance and size. Proc. Natl. Acad. Sci. **99**(20), 12917–12922 (2002)
4. Freeman, S.C., Freeman, L.C.: The Networkers Network: A Study of the Impact of a New Communications Medium on Sociometric Structure. Social sciences research reports. School of Social Sciences University of Calif (1979)
5. Kaiser, M., Hilgetag, C.C.: Nonoptimal component placement, but short processing paths, due to long-distance projections in neural systems. PLoS Comput. Biol. **2**(7), e95 (2006)
6. Leskovec, J., Adamic, L.A., Huberman, B.A.: The dynamics of viral marketing. ACM Trans. Web (TWEB) **1**(1), 5 (2007)
7. Leskovec, J., Huttenlocher, D., Kleinberg, J.: Signed networks in social media. In: Proceedings of the SIGCHI Conference on Human Factors in Computing Systems, pp. 1361–1370. ACM (2010)
8. Leskovec, J., Kleinberg, J., Faloutsos, C.: Graphs over time: densification laws, shrinking diameters and possible explanations. In: Proceedings of the Eleventh ACM SIGKDD International Conference on Knowledge Discovery in Data Mining, pp. 177–187. ACM (2005)
9. Leskovec, J., Kleinberg, J., Faloutsos, C.: Graph evolution: densification and shrinking diameters. ACM Trans. Knowl. Discov. Data (TKDD) **1**(1), 2 (2007b)
10. Leskovec, J., Lang, K.J., Dasgupta, A., Mahoney, M.W.: Community structure in large networks: natural cluster sizes and the absence of large well-defined clusters. Internet Math. **6**(1), 29–123 (2009)
11. Liu, Y-Y., Slotine, J-J., Barabási, A-L.: Observability of complex systems. Proc. Natl. Acad. Sci. **110**(7), 2460–2465 (2013)
12. Lusseau, D., Schneider, K., Boisseau, O.J., Haase, P., Slooten, E., Dawson, S.M.: The bottlenose dolphin community of doubtful sound features a large proportion of long-lasting associations. Behav. Ecol. Sociobiol. **54**(4), 396–405 (2003)
13. Milo, R., Itzkovitz, S., Kashtan, N., Levitt, R., Shen-Orr, S., Ayzenshtat, I., Sheffer, M., Alon, U.: Superfamilies of evolved and designed networks. Science **303**(5663), 1538–1542 (2004)
14. Milo, R., Shen-Orr, S., Itzkovitz, S., Kashtan, N., Chklovskii, D., Alon, U.: Network motifs: simple building blocks of complex networks. Science **298**(5594), 824–827 (2002)
15. Penzl, T.: Algorithms for model reduction of large dynamical systems. Linear Algebra Appl. **415**(2), 322–343 (2006)
16. Richardson, M., Agrawal, R., Domingos, P.: Trust management for the semantic web. In: The Semantic Web-ISWC 2003, pp. 351–368. Springer (2003)

17. Ripeanu, M., Foster, I.: Mapping the gnutella network: Macroscopic properties of large-scale peer-to-peer systems. In: Peer-to-Peer Systems, pp. 85–93. Springer (2002)
18. Shen-Orr, S.S., Milo, R., Mangan, S., Alon, U.: Network motifs in the transcriptional regulation network of escherichia coli. Nat. Genetics **31**(1), 64–68 (2002)
19. Takac, L., Zabovsky, M.: Data analysis in public social networks. In: International Scientific Conference and International Workshop Present Day Trends of Innovations, pp. 1–6 (2012)
20. Watts, D.J., Strogatz, S.H.: Collective dynamics of 'small-world' networks. Nature **393**(6684), 440–442 (1998)
21. Young, M.P.: The organization of neural systems in the primate cerebral cortex. Proc. R. Soc. Lond. B: Biol. Sci. **252**(1333), 13–18 (1993)

Appendix B
Reduction of Relative Degree by Additional Drivers and Sensors

This chapter contains additional information for Chap. 3.

B.1 Details of the Case Studies

The Modified Nodal Analysis Problem

In the MNA problem, voltage sources are connected to the port of multiport to determine its the admittance matrix [2]. A simple two-port RLC circuit example can be seen in Fig. B.1.

With the voltages (u) and currents (i), the multiport can be represented with the MNA equations, Eqs. (B.1) and (B.2).

$$\mathbf{C}\dot{x} = \mathbf{G}x + \mathbf{B}u$$
$$i = \mathbf{L}^{\mathrm{T}}x \qquad \text{(B.1)}$$

Fig. B.1 A simple two-port RLC circuit

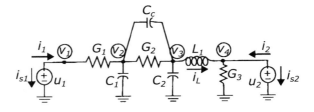

Fig. B.2 Network representation of the simple two-port RLC circuit

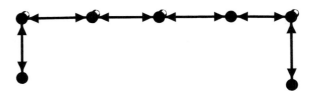

Fig. B.3 The smallest PEEC problem for two planes [3]

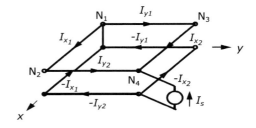

$$
\begin{bmatrix}
0 & 0 & 0 & 0 & 0 & 0 & 0 \\
0 & (C_c + C_1) & -C_c & 0 & 0 & 0 & 0 \\
0 & -C_c & (C_2 + C_c) & 0 & 0 & 0 & 0 \\
0 & 0 & 0 & 0 & L_1 & 0 & 0 & 0 \\
0 & 0 & 0 & 0 & 0 & 0 & 0 \\
0 & 0 & 0 & 0 & 0 & 0 & 0 \\
0 & 0 & 0 & 0 & 0 & 0 & 0
\end{bmatrix}
\begin{bmatrix}
\dot{v}_1 \\ \dot{v}_2 \\ \dot{v}_3 \\ \dot{v}_4 \\ \dot{i}_L \\ \dot{i}_{s1} \\ \dot{i}_{s2}
\end{bmatrix}
= -
\begin{bmatrix}
G_1 & -G_1 & 0 & 0 & 0 & 1 & 0 \\
-G_1 & (G_1 + G_2) & -G_2 & 0 & 0 & 0 & 0 \\
0 & -G_2 & G_2 & 0 & 1 & 0 & 0 \\
0 & 0 & 0 & G_3 & -1 & 0 & 1 \\
0 & 0 & -1 & 1 & 0 & 0 & 0 \\
-1 & 0 & 0 & 0 & 0 & 0 & 0 \\
0 & 0 & 0 & -1 & 0 & 0 & 0
\end{bmatrix}
\begin{bmatrix}
v_1 \\ v_2 \\ v_3 \\ v_4 \\ i_L \\ i_{s1} \\ i_{s2}
\end{bmatrix}
+
\begin{bmatrix}
0 & 0 \\ 0 & 0 \\ 0 & 0 \\ 0 & 0 \\ 0 & 0 \\ -1 & 0 \\ 0 & -1
\end{bmatrix}
\begin{bmatrix}
u_1 \\ u_2
\end{bmatrix}
$$

$$
\begin{bmatrix} i_1 \\ i_2 \end{bmatrix} =
\begin{bmatrix}
0 & 0 & 0 & 0 & 0 & -1 & 0 \\
0 & 0 & 0 & 0 & 0 & 0 & -1
\end{bmatrix}
\begin{bmatrix}
v_1 \\ v_2 \\ v_3 \\ v_4 \\ i_L \\ i_{s1} \\ i_{s2}
\end{bmatrix}
$$

(B.2)

The proposed method utilizes the network representation of this state-transition matrix depicted in Fig. B.2.

This simple example demonstrated only the basic building blocks of the MNA problems. The structures of the studied more complex MNA problems will be presented in the following section (on Fig. B.6 and Fig. B.7).

The Partial Element Equivalent Circuit Problem

The smallest possible example for the PEEC problem can be seen in Fig. B.3.

As can be seen, PEEC is used to model 3-dimensional interconnects. Partial inductance included by each of the connection. If node N_1 is shorted and into node N_4 current is injected, then the self- inductance of N_4 is $L_{44} = \frac{V_4}{sI}$ [1]. The circuit equations can be given as in Eq. (B.3).

$$
\begin{bmatrix}
0 & 0 & 0 & 0 & 1 & 0 & 1 & 0 & 1 \\
0 & 0 & 0 & 0 & -1 & 0 & 0 & 1 & 0 \\
0 & 0 & 0 & 0 & 0 & 1 & -1 & 0 & 0 \\
0 & 0 & 0 & 0 & 0 & -1 & 0 & -1 & 0 \\
1 & -1 & 0 & 0 & -Lx_{11} & -Lx_{12} & 0 & 0 & 0 \\
0 & 0 & 1 & -1 & -Lx_{12} & -Lx_{22} & 0 & 0 & 0 \\
1 & 0 & -1 & 0 & 0 & 0 & -Ly_{11} & -Ly_{12} & 0 \\
0 & 1 & 0 & -1 & 0 & 0 & -Ly_{12} & -Ly_{22} & 0 \\
1 & 0 & 0 & 0 & 0 & 0 & 0 & 0 & 0
\end{bmatrix}
\begin{bmatrix}
V_{N1} \\ V_{N2} \\ V_{N3} \\ V_{N4} \\ sI_{x1} \\ sI_{x2} \\ sI_{y1} \\ sI_{y2} \\ sI_{sh}
\end{bmatrix}
=
\begin{bmatrix}
0 \\ 0 \\ 0 \\ I_s \\ 0 \\ 0 \\ 0 \\ 0 \\ 0
\end{bmatrix}
$$

(B.3)

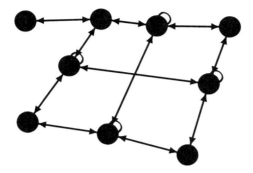

Fig. B.4 Network representation of the example problem for two planes

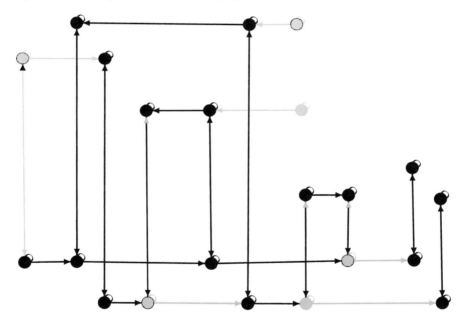

Fig. B.5 Network representation of the HEN case study at $K^+ = 3$ for both the mCLASA and GDFCMSA method. The turquoise dots denote the fixed sensors (S_f) while the green dots denote the placed additional sensors (S_c) determined by mCLASA

The network representation of the matrix of of state variables is depicted in Fig. B.4.

This simple example demonstrated only the basic building blocks of the PEEC problem. The structure of the studied more complex problem will be presented in the following section (on Fig. B.8).

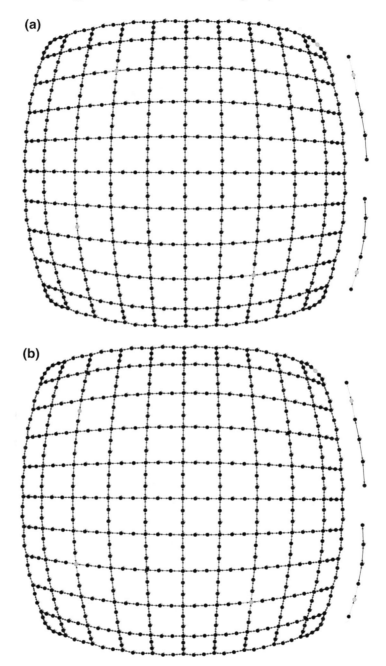

Fig. B.6 Network representation of the the MNA_1 case study at $K^+ = 3$ obtained by **a** the mCLASA and **b** the GDFCMSA methods. The turquoise dots denote the fixed sensors (S_f) while the green dots denote the optimal candidate sensors (S_c) determined by mCLASA

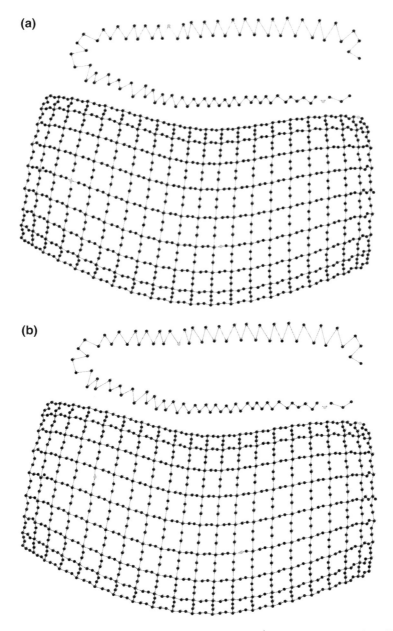

Fig. B.7 Network representation of the MNA_4 case study at $K^+ = 3$ obtained by (**a**) the mCLASA and (**b**) the GDFCMSA methods. The turquoise dots denote the fixed sensors (S_f) while green dots denote the optimal candidate sensors (S_c) determined by mCLASA

(a)

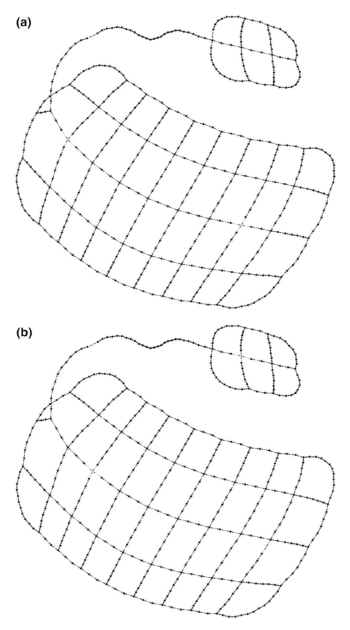

(b)

Fig. B.8 Network representation of the PEEC case study at $K^+ = 3$ obtained by **a** the mCLASA and **b** the GDFCMSA methods. The turquoise dots denote the fixed sensors (S_f) while green dots denote the optimal candidate sensors (S_c) determined by mCLASA

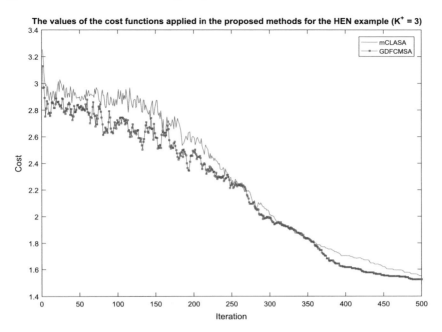

Fig. B.9 Convergence of the proposed methods in the case of case study HEN. The colour red denotes the method mCLASA while the colour blue denotes the method GDFCMSA

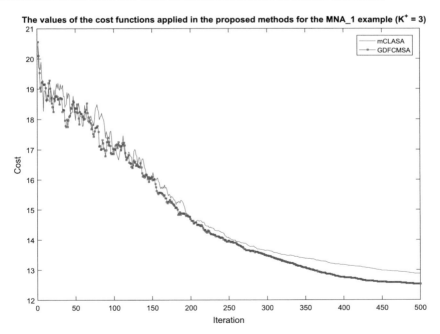

Fig. B.10 Convergence of the proposed methods in the case of case study MNA_1. The colour red denotes the method mCLASA while the colour blue denotes the method GDFCMSA

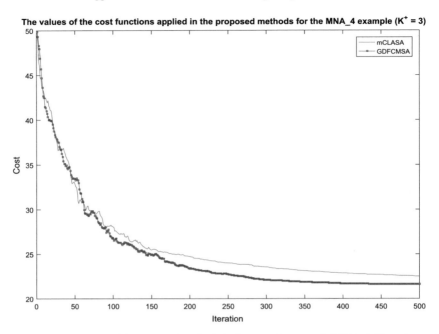

Fig. B.11 Convergence of the proposed methods in the case of case study MNA_4. The colour red denotes the method mCLASA while the colour blue denotes the method GDFCMSA

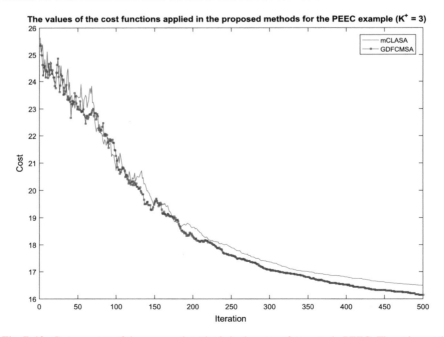

Fig. B.12 Convergence of the proposed methods in the case of case study PEEC. The colour red denotes the method mCLASA while the colour blue denotes the method GDFCMSA

B.2   Network Representations of the Case Studies

The network representations contain the fixed (S_f) and candidate (S_c) sensors noted with turquoise and green, respectively. As the results of the mCLASA and GDFCMSA methods can differ, both results are presented. The only exception is the HEN case study, as the two methods provide identical solutions. In the following, Fig. B.5, B.6, B.7 and B.8 show the solutions provided by the two methods for the HEN, MNA_1, MNA_4 and PEEC case studies.

B.3   Convergence of the Solutions

The convergence of the HEN case study was introduced in Sect. 3.4.3. Here, the convergence of the remaining three case studies are also presented in Figs. B.9, B.10, B.11, and B.12.

References

1. Heeb, H., Ruehli, A.E., Eric Bracken, J., Rohrer, R.A.: Three dimensional circuit oriented electromagnetic modeling for VLSI interconnects. In: Proceedings, IEEE 1992 International Conference on Computer Design: VLSI in Computers and Processors, 1992. ICCD'92, pp. 218–221. IEEE (1992)
2. Odabasioglu, A., Celik, M., Pileggi, L.T.: Prima: passive reduced-order interconnect macromodeling algorithm. In: Proceedings of the 1997 IEEE/ACM International Conference on Computer-Aided Design, pp. 58–65. IEEE Computer Society (1997)
3. Zhou, F., Ruehli, A.E., Fan, J.: Efficient mid-frequency plane inductance computation. In: 2010 IEEE International Symposium on Electromagnetic Compatibility (EMC), pp. 831–836. IEEE (2010)

Index

© The Author(s), under exclusive license to Springer Nature Switzerland AG 2020
D. Leitold et al., *Network-Based Analysis of Dynamical Systems*,
SpringerBriefs in Computer Science,
https://doi.org/10.1007/978-3-030-36472-4

Printed in the United States
By Bookmasters